Radiography in Veterinary Practice at a Glance (Including Diagnostic Imaging Techniques)

ABOUT THE AUTHORS

Dr. A.K. Gangwar, a graduate from the College of Veterinary Science & Animal Husbandry, Mathura and got ICAR-JRF for completing M.V.Sc. in Veterinary Surgery and Radiology from Indian Veterinary Research Institute, Izatnagar. Dr. Gangwar joined College of Veterinary Science & Animal Husbandry, N.D. University of Agriculture & Technology, Kumarganj, Faizabad (UP) in the year 2002 and established the Department of Veterinary Surgery & Radiology. Presently Dr. Gangwar is working as Associate Professor in the Department of Veterinary Surgery & Radiology of the same college. Dr. Gangwar is working on Biomaterials and Regenerative medicine in veterinary practice. Dr. Gangwar published a textbook, more than 70 research/clinical articles in different peer reviewed international and national journals and presented more than 25 scientific papers in international and national seminars.

Dr. Khangembam Sangeeta Devi, completed M.V.Sc. in Veterinary Surgery and Radiology from Ranchi Veterinary College, Ranchi. Dr. Sangeeta joined College of Veterinary Science & Animal Husbandry, N.D. University of Agriculture & Technology, Kumarganj, Faizabad (UP) as Teaching Associate in the year 2005 and from 2012 onwards she holds the position of Assistant Professor in the Department of Veterinary Surgery & Radiology of the same college. Dr Sangeeta published a textbook, more than 40 research/clinical articles in different peer reviewed international and national journals and presented more than 15 papers in international and national seminars.

Dr. Naveen Kumar joined Indian Veterinary Research Institute as a Scientist in the year 1989 and from 2006 onwards he holds the position of Principal Scientist. Dr. Kumar has more than 25 years of research experience in the field of Regenerative Medicine and Biomaterials in Veterinary Practice. Dr. Kumar is a Fellow of National Academy of Veterinary Science, which is given by the apex premier society, National Academy of Veterinary Science, for the development of Veterinary Sciences in India, for outstanding contribution in the field of Veterinary Science.

Radiography in Veterinary Practice at a Glance

(Including Diagnostic Imaging Techniques)

A. K. Gangwar
Khangembam Sangeeta Devi
Naveen Kumar

2015

Daya Publishing House®

A Division of

Astral International Pvt. Ltd.

New Delhi – 110 002

© 2015 AUTHORS
ISBN 9789351305408

Published by	:	**Daya Publishing House®**
		A Division of
		Astral International Pvt. Ltd.
		– ISO 9001:2008 Certified Company –
		House No. 96, Gali No. 6,
		Block-C, 30ft Road, Tomar Colony, Burari
		New Delhi-110 084
		E-mail: info@astralint.com
		Website: www.astralint.com
Sales Office	:	4760-61/23, Ansari Road, Darya Ganj
		New Delhi-110 002 Ph. 011-23245578, 23244987
Laser Typesetting	:	**Classic Computer Services**, Delhi - 110 035
Printed at	:	**Replika Press Pvt. Ltd.**

PRINTED IN INDIA

PREFACE

The goal of thé authors is to provide latest information of radiology, radiotherapy and modern diagnostic imaging techniques to undergraduate and postgraduate students, professors, field veterinarians and radiographers who are actively involved in radiography of the animais. Although the book has been framed mainly as per syllabus approved by Veterinary Council of India but it is of great use for postgraduate students, professors involved in radiology and radiotherapy, veterinary surgeons, radiographers, radiologists, field veterinarians and researchers. The book is relevant, concise and easy to read describing general principles of radiology, radiotherapy and modern diagnostic techniques. The authors feel a great privileged in bringing out the first edition of the book entitled "Radiography in Veterinary Practice at a Glance (Including Diagnostic Imaging Techniques)". The book has been prepared by consulting several standard textbooks and journals of related field. A number of illustrations and more than 100 good quality photographs of positioning of animals, normal and contrast radiographs of different body parts and radiographs of diseases in animais hâve been incorporated at places to make the text more meaningful. We are indebted to Dr. A.K. Sharma, Dr. Adarsh Kumar, Dr. J. Mohindroo, Dr. J.Y. Waghaye, Dr N.K. Singh, Dr. M. Hoque, Dr. R. P. Pandey, Dr. S.K. Tiwary, Dr. A.K. Das, Dr. R. B. Kushwaha, Dr. R. N. Chaudhary, Dr. J.K. Das, Dr. S.P. Tyagi, Dr. Kiranjeet, Dr. Ashwathy, Dr. S.K. Maiti, Dr. Vineet Kumar, Dr. Ramesh Tiwary, Dr. G.D. Singh, Dr. Arvind Sharma, Dr. Himanshu Singh, Dr. Rahul Udehiya, Dr. Surbhi Gupta, Dr. Jasmeet Singh, Dr. Rukmani Dewangan, Dr. Anil Bishnoi, Dr. A. K. Majhi, Dr. Samar Halder, Dr. Amit Bisla. , Dr. Dayamon D. Mathew, Dr. Warson, Dr. Irawati Sarode and Dr. Kaarthick, D.T., who have made helpful suggestions. We also appreciate our students, Dr. Ajeet Kumar Singh, Dr. Ghanshyam Patel, Dr. Nitesh Katiyar, Dr. Nishant Yadav and Dr.S.S.Kale who hâve directly or indirectly helped with this project. Finally, Dr. A.K. Gangwar and Dr. Kh. Sangeeta Devi are exceedingly grateful to our lovely child (Ansi and Sanskriti) for their support, patience, and encouragement. We shall be grateful to the professionals and colleagues for thé constructive suggestions for the betterment of this book. Authors are hopeful that this book will serve the intended goal.

Dr. A. K. Gangwar
Dr. Kh. Sangeeta Devi
Dr. Naveen Kumar

Contents

[viii]

1 Timeline of Radiology and Other Diagnostic Imaging Techniques

TIME LINE

November 8, 1895	Wilhelm Conrad Roentgen	Discovered electromagnetic radiation, which he called X-rays, as he didn't know the nature of these rays.
1896	Weker	Develops the first skull radiograph
	Konig and Morten	Developed the first dental radiograph
	Trowbridge	Developed first oil immersed X-ray tube
	Pupin	Made first intensifying screen
	Wright	Developed photographic paper for recording of X-ray image
	Paton and Duncan	Published first veterinary radiograph of equine foot
	R. Eberlin,	**Father of Veterinary Radiology** Use of X-ray in veterinary practice
1898	Cannon	Contrast studies of GIT using Bismuth
1901	Wilhelm Conrad Roentgen	First Nobel Prize in Physics
1902	G. Holtz Knecht	Developed first dosimeter for radiation therapy
1905		First roentgenological congress held at Berlin under the chairmanship of R. Eberlin, a Veterinarian
1913	Coolidge	Introduces cathode tube/ high vacuum X-ray tubes
	Gustav P. Bucky	Invented grid to check the scatter radiation
1914	W.H. Bragg and W.L. Bragg	Discovered that X-rays could be reflected
1916	Wilhelm Trendelenburg	Developed red adaptation goggles
1918	Walter Dandy	Develops ventriculography
	Eastman	Introduces radiographic film
		Line focus principle was discovered
1920		Double coated X-ray films were developed
		Use of iodine based contrast agents started
	Dr. Hollis potter	Discovered moving grid
1921	Andre Bocage	Creates the first tomography (body-section imaging)
1927	Moniz	Cerebral Angiography using contrast media
1928		International recommendations on radiation safety precautions were published
1930	C.C. Lauriston	Developed supervoltage single section X-ray tube
1934	Joliot and Curie	Discover artificial radionuclides
1935	Ziedes de Plantes	Developed substraction technique
1937	Chester F. Carlson	Invented Xeroradiography
1942	Karl Theodore Dussik	Published the first paper on medical ultrasonics

Contd...

Contd...

1945	Gray Schnelle	Wrote first American Book on Veterinary Radiology
1946	Schoenander	Develops the film cassette changer which allowed a series of cassettes to be exposed at the rate of 1.5 cassettes per second
1950's		Development of the image intensifier and X-ray television
1951	Benedict Cassen	Scintillation camera
1954	Hertz and Elder	Echocardiographic
1957	Ian Donald	• Developed the first contact scanner • First ultrasound of the uterus during pregnancy
1958	Hal Anger	Gamma camera with the technetium radioisotope
	Alois Pommer	Published treatise on Veterinary Radiotherapy
1960		Use of first radiographic film with polyester base
	American Veterinary Radiological Society	Started Journal Veterinary Radiology. Now Veterinary Radiology and Ultrasound
1962	Kuhl	Developed single positron emission computed tomography (SPECT)
1972	Godfrey Hounsfield and Allan McLeod Cormack	Invented X-Ray Computed tomography. Both scientists were rewarded Nobel Prize in Physiology or Medicine in 1979
	Paul Lauterbur	Magnetic Resonance Imaging (MRI)
	Pupin	Rare earth intensifying screen developed
1975	M. M. Ter-Pogossian, and M. E. Phelps	Positron Emission Tomography (PET)
1977	Raymond Damadian	Designs and invents the first MRI scanner
1980's	Fuji	Develops CR technology
1987	Charles Dumoulin	MRA (Magnetic Resonance Angiography)
1992		International radiology Association was changed from Veterinary Radiology to Veterinary Radiology and Ultrasound
1993		Functional MRI (fMRI).

RADIOGRAPHY

Making a radiographic record of internal structures of the body by exposure of the film by X-rays is known as radiography.

Types of Radiography

1. **Spot film radiography:** Making of localized instantaneous radiographic exposures during fluoroscopy.

2. **Stress radiography:**

    • Stress (traction, rotation or wedge forces) is placed on structures being radiographed.

    • Most commonly used in the diagnosis of joints and spinal disorders such as lumbo-sacral instability, wobbler syndrome and atlanto-axial instability.

3. **Serial radiography:** Making of more than one exposure of a particular area at different intervals.

4. **Mucosal relief radiography:** Performed for revealing any abnormality of the intestinal mucosa.

5. **Intra-oral radiography:**
   - Small non-screen film is placed in the mouth and x-rays are directed from outside the mouth.
   - Used in dental radiography.

6. **Body section radiography:** e.g. Laminography or tomography.

7. **Contrast radiography:** Radiograph is taken after introduction of contrast agents.

8. **Xeroradiography:** The process of making a type of dry x-rays in which a picture of the body is recorded on paper rather than on the film.

9. **Interventional radiography:** In Interventional radiography the medical procedures (usually minimally invasive) are performed with the guidance of imaging technologies.

10. **Teleradiology:** Teleradiology is the transmission of radiographic images from one location to another for interpretation by a radiologist.
    - X-rays were discovered by Wilhelm Conrad Roentgen on November 8, 1895 and got the first Nobel Prize for physics in 1901.

Radiology (Roentgenology): It is the branch of medical/veterinary science that employs the use of imaging for diagnostic and therapeutic purpose.

Radiologist: Any person qualified in veterinary/medical science and radiological physics who uses an array of imaging technologies such as X-ray radiography, ultrasound, computed tomography (CT), nuclear medicine, positron emission tomography (PET) and magnetic resonance imaging (MRI) to diagnose or treat diseases.

Radiographer or radiologic technologist: The acquisition of imaging is usually carried out by the radiographer or radiologic technologist. The radiologist then interprets or "reads" the images and produces a report of their findings.

Radiograph / roentgenograph / skiagram: Radiograph is produced on silver-impregnated films by transmitting X-rays through the patient. It is the photographic record of the extent of penetrability of X-rays through the exposed tissue part.

Radiation

RADIOLOGY SECTION

The overall set up of radiology section should consider the following requirements

1. Space
2. X-ray machines
3. Electric supply
4. Accessory equipments

1. SPACE

The site for construction of building for installation of X-ray unit should be located

A. Within the premises of clinic and surgery area and from which all unnecessary persons could be excluded when the x-ray unit is under operation to minimize radiation hazards.

B. Where the animals to be radiographed can easily be brought to the exposure room.

The area should have space for-

(a) X-ray room or exposure room
(b) Control panel area
(c) Dark room
(d) Radiologist and radiographer room
(e) Interpretation room
(f) Store room for keeping unexposed films and other accessories
(g) Waiting area
(h) Film file room
(i) Teaching hall

(a) **X-ray room or exposure room**

- It should have large space which reduces radiation exposure of personnel due to decreased scatter radiation.
- It should have restraining devices like travis, casting trolley etc.
- The floor should not be slippery.
- The windows and doors should be covered with thick curtain.

- Thick concrete walls (15 cm thick) should be painted with white lead paint.

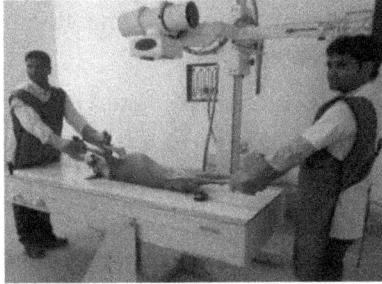

Fig. 2.1: Posture of persons for positioning of animal during exposure

(b) Control panel area

- The area should be located in such a way that it is seldom in line with the primary X-ray beam. There should be a protective partition with a lead glass window (30x30cm) to view animal and machine during exposure (**Fig. 2.2**).

Fig. 2.2: Handling of control panel

(c) Dark room

- The dark room should be adjacent to the X-ray room and must be light proof
- It should have well demarcated dry bench and wet bench in orderly sequence of successive stage of work. Each bench should be fitted with safe light (low watt frosted bulb, maximum 10 watt, covered by specific filters) at minimum height of 3 feet from working table.
- There should be provision for sufficient running water.
- It should have a lead box to store unexposed films currently in use.
- Room temperature should not be too cold or too hot.
- It should be fitted with film drying rack.

(d) **Radiologist and radiographer room**
- It should be located near the X-ray room/dark room.
- It should be fitted with viewing illuminator.

(e) **Interpretation room**
- A separate interpretation room should be planned
- Provision should be made for at least two viewing illuminator.

(f) **Store room for keeping unexposed films and other accessories**
- The unexposed sealed packets of X-ray films should be stored away from source of radiation and other accessories can be stored in this room.
- If separate room is not feasible then a portion of dark room can be used for this purpose.

(g) **Waiting area**
- Waiting area should be located outside the x-ray room.
- The area should have minimum radiation hazards and minimum obstruction in the flow of work.

(h) **Film file room**
- It should be near the radiologist's office.
- The room should have sufficient provision of storing racks or cabinets for film boxes so that film can be systematically stored year wise, species wise or as in required manner.

(i) **Teaching hall**
- In teaching institutes, there should be provision for teaching hall for teaching purpose.
- It should be fitted with viewing illuminators.

2. X-RAY MACHINES

X-ray machines are main source of X-ray generation and can be grouped into three main categories:

(i) Portable X-ray Machines
- Easy to transport.
- Maximum output usually varies from 70-110 kV and 15-35mA.

(ii) Mobile X-ray Machines
- Most machines are movable on wheels or on smooth surface within the radiology section.
- Maximum output usually varies from 90-125 kV and 40-300mA.

(iii) Fixed X-ray machines

- These machines are mostly ceiling mounted having telescoping tube.
- Maximum output usually varies from 120-200 kV and 300-1000mA.

3. ACCESSORIES

(i) **X-ray film clip:** These are ordinary stainless steel clips for hanging the processed X-ray film **(Fig. 2.3a)**.

(ii) **Collimators:** They are made from lead and used to limit size of the primary X-ray beam. Commonly used collimators are diaphragm, cone **(Fig. 2.3b)** and cylinder **(Fig. 2.3c)** and used at the window of the X-ray tube. The collimator reduces scatter radiation and patient radiation dose is decreased.

Fig. 2.3a: X-ray film clips

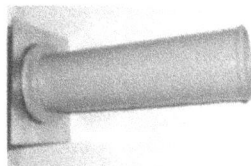

Fig. 2.3b: Collimator (Cyllimator Cylinder)

Fig. 2.3c: Colliomator (Cone)

(iii) **Cassette or film holder:** It is a light proof box designed to hold X-ray film for taking an X-ray exposure **(Fig. 2.3d)**.

(iv) **Film drying rack:** film hangers are loaded on this rack for drying of exposed films **(Fig. 2.3e)**.

(v) **Film markers:** These are letters and numbers of lead and used for permanent marking on X-ray film **(Fig. 2.3f)**.

(vi) **Contrast scale:** An instrument to find out a wide range and great number of shades of gray with little difference in the adjacent tones of a radiographic image **(Fig. 2.3g)**.

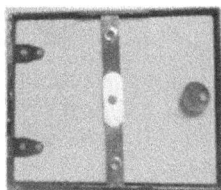

Fig. 2.3d: Cassette or film holder

Fig. 2.3e: Film drying rack

Fig. 2.3f: Film markers

(vii) **Lead gloves and aprons:** During exposure they protect the front portion of the body and hands. They should have 0.25 mm and 0.5 mm lead equivalent, respectively. These items should be kept on a roller slant to prevent it from damage **(Fig. 2.3h and 2.3j)**.

(viii) **Caliper:** It is used to measure the thickness of the object to be radiographed **(Fig. 2.3i).**

Fig. 2.3g: Constrast scale   **Fig. 2.3h:** Lead glove   **Fig. 2.3i:** Caliper

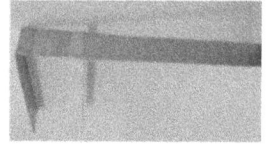

(ix)   **Film hangers:** Exposed films are loaded in film hangers for easy processing. Following types of hangers are available- Clip type, channel type and tension type **(Fig. 2.3k).**

(x)   **Viewing illuminator:** It is used for proper radiographic interpretation **(Fig. 2.3l).**

Fig. 2.3j: Lead apron   **Fig. 2.3k:** Film hanger   **Fig. 2.3l:** Viewing illuminator

(xi)   **Safe light:** It is a box containing a low watt (10 watt maximum) frosted bulb covered by a specific filter **(Fig. 2.3m).**

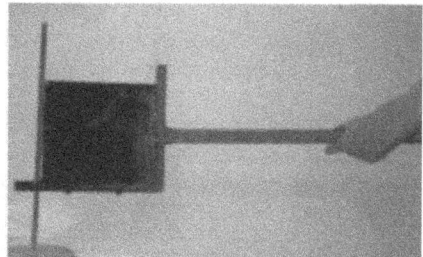

(xii) **Cassette holder:** It is used for holding cassette during radiography of large animals in standing position **(Fig. 2.3n).**

Fig. 2.3m: Safe light   **Fig. 2.3n:** Cassette holder

(xiii) **Blocks:** These are made of wood and used for positioning of part/ to elevate animal's foot to the centre of X-ray beam. Rickman navicular block is used for radiography of navicular bone of horse.

(xiv) **Cassette pass box:** It is a device fitted in the wall between exposure room and dark room. The exposed cassettes are sent to the dark room through cassette pass box from the exposure room.

(xv)   **Developing thermometers and heaters:** Required to keep the temperature of the processing chemicals.

(xvi)   **Exposure table:** It is used to control and position small animals during exposure.

(xvii)   **Film cutter/ film trimmer:** It is used to cut the X-ray film.

(xviii)   **Film divider:** It is a sheet of lead rubber or lead metal and used to divide film for getting two or more than two exposures on the same film.

(xix)   **Film drier:** It is used to dry films after processing.

(xx)   **Film storage cabinet / box:** It is a wooden or metallic box with a lead lining and used to store unexposed films.

(xxi)   **Filters:** These are aluminum sheets of 1 mm thickness. It is placed over the window to filter out useless, soft non penetrating X-rays from the primary X-ray beam.

(xxii)   **Floating thermometer:** It is used to measure temperature of processing solution before film processing.

(xxiii)   **Fluoroscopic screen:** It is used to view an X-ray image of moving body parts.

(xxiv)   **Goggles with red glasses:** For dark adaptation while using the florescent screening.

(xxv)   **Gonad shield:** Available in three different sizes of large, medium and small and having lead equivalent of 0.5 mm.

(xxvi)   **Grid:** It is used to absorb secondary radiation and p[laced between the cassette and part of the animal. It is made up of sheets of lead stripes laid side by side with radiolucent space which is packed with aluminum or wood.

(xxvii)   **Intensifying screens:** Used to reduce the amount of X-rays necessary for exposure and thus reduce the time required to expose an X-ray film.

(xxviii)   **Lead goggle:** It is used to protect the eyes from radiation.

(xxix)   **Lead screen:** It is used as a barrier between source of radiation and individual.

(xxx)   **Measuring tape and Caliper:** Required to measure the size and thickness of the part to be radiographed.

(xxxi)   **Potter-bucky diaphragm:** It is a type of grid that moves during radiographic exposure. Grid lines are not produced by the use of moving grid as seen in case of stationary grid.

(xxxii)   **Radiation safety monitoring device:** These devices are used to monitor level of radiation around the radiological unit and the staff working in radiation field.

(xxxiii) **Screen cleaning solution:** The screen should be cleaned regularly with this solution.

(xxxiv) **Timer:** It is used to fix time for various processing steps.

(xxxv) **TLD badges:** Thermo luminescence badges are necessary for recording the radiation dosages to the radiographer and attendant.

(xxxvi) **X-ray films**

    A. **Types:** Screen type, non screen type, automatic processor film and occlusal films.

    B. **Sizes (in inches):** 5x7, 6.5x8.5, 6x12, 6x15, 8x10, 10x12, 11x14, 12x12, 12x15, 14x14 and 14x17.

(xxxvii) **Stirrer (stainless steel):** Separate stirrer should be used for mixing of developer and fixer.

3   Role and Usage of Sedatives/ Anesthetics for Radiography

Radiography is one of the most commonly used diagnostic tools in veterinary practice. It provides a large amount of information to the veterinarian by noninvasive means. It does not alter the disease process or cause unacceptable discomfort to the animal. Patient restraining is very important to obtain a quality radiograph. Animals must be adequately restrained and positioned to obtain quality radiographic images. Due to ionizing radiation regulations, the staff should not be allowed to hold the animal without radiation safety devices. The personnel with appropriate protective apparel may manually restrain animals; however, manual restraint should be kept to a minimum. In some countries, manual restraining of animals is not allowed except under explicitly defined circumstances. Sedation or short-acting anesthesia is often necessary. Chemical restraint lessens the need for manual restraint, which leads to fewer poor or unacceptable radiographs and usually shortens the time required to complete the examination. In many instances, animals can be restrained using sandbags, tape, and foam pads. It is preferable to chemically immobilize the animal as long as there is not a medical contraindication.

Although radiography itself is painless, sedation is often desirable in order to reduce anxiety and stress associated with the procedure, as well as to control pain associated with manipulation of animals with painful disorders such as fractures and arthritis. Sedation or general anesthesia is recommended in most cases to achieve accurate positioning, to reduce the number of retakes and to ensure standard positioning which will ease interpretation. It can be useful to add analgesia to the sedation to minimize any discomfort that might be caused by the patient's condition.

Administer a sedative or general anesthetic to get the best possible picture without causing any unnecessary distress to the pet or any risk to the staff. Use the minimum amount of sedation as needed for the particular X-ray.

When taking radiographs of the different body parts specially hips, skull, oral cavity, and spine, the animal must remain perfectly motionless to obtain quality radiographs. Animal motion may be minimized by decreasing exposure time and maximizing mA to achieve the required mAs for the body region examined. Other technical adjustments, such as increasing the kVp or shortening the film focus distance, may be made in some cases. However, major changes in film focus distance"will likely cause serious degradation of the image. Sometimes, the condition is painful, or the positioning is uncomfortable. For these reasons animals are usually sedated, or lightly anesthetized when these types of radiographs are taken. It makes the procedure less uncomfortable for the animal, and allows getting a good radiograph the first time. That way the radiographer does not have to put the animal through the process multiple times before he get a radiograph that is acceptable.

Digital Radiography is an excellent diagnostic tool. It allows us to view inside the body to detect internal conditions, abnormal fluids, broken bones, and foreign body obstructions. Usually this procedure can be performed without sedation. However, if a patient is experiencing pain or is fractious, we can safely administer a pain medication or sedatives before taking the radiographs

SEDATION AND/OR ANESTHESIA FOR RADIOGRAPHY

The physical status of the animal at the time of presentation depends on whether or not to use sedation and/or anesthesia for diagnostic orthopedic radiographs. Anesthetics and the sedatives used for its maintenance in general, the potential reasons for using an anesthetic protocol for the pets when undergoing diagnostic and/or emergency radiographs for various conditions, as well as the use of anesthesia.

Anesthetic drugs may be grouped into categories by their mode of action and/or their purpose in an -anesthetic protocol. There are those used as premedicants for general anesthesia, those used to induce general anesthesia, those used to maintain anesthesia, and those used for pain relief alone (analgesics). Drugs are selected based upon the health status of the patient, expected duration of the procedure, and the degree of pain anticipated. Premedicating the patient, pre-oxygenating prior to induction, and maintaining an intravenous catheter and administering IV fluids during general anesthesia can improve the relative safety of any drug or drug combination.

Invasive diagnostic and therapeutic interventional radiological procedures can be painful and anxiety provoking. The combination of propofol and ketamine may minimize the need for supplemental opioid analgesics and has the potential to provide better sedation with less toxicity than either drug alone. The combination of propofol and ketamine (propofol 0.5 mg/kg + ketamine 0.5 mg/kg, and propofol 0.5 mg/kg + ketamine 0.25 mg/kg intravenously) have been used for sedation during interventional radiological procedures, showed no clinically significant hemodynamic changes or side effects and both appeared to prompt early recovery time. However, the anesthesiologist recommends propofol 0.5 mg/kg + ketamine 0.5 mg/kg for sedation and analgesia during interventional radiological procedures, rather than propofol 0.5 mg/kg + ketamine 0.25 mg/kg because the former combination is associated with reduced propofol requirements and therefore less oversedation.

PREMEDICANTS USED FOR RADIOGRAPHY

Premedicants are used to reduce anxiety, provide pre-emptive analgesia, reduce the amount of anesthetic induction and maintenance anesthesia required, produce some muscle relaxation, and help provide "smooth" inductions and recoveries. Preanesthetic drugs with analgesic effects, such as sedatives (opioids and non-opioids), tranquilizers (Phenothiazine derivatives, butyrophenone derivatives and benzodiazapines) and neuroleptanalgesics are often included in a premedication scenario. However, premedicants may include agents that have no analgesic or sedative qualities at all, such as anticholinergics (Atropine, glycopyrrolate) but used for other purposes, such as their cardiovascular protective effects. These drugs are usually administered as a subcutaneous (under the skin) or intramuscular injection around 15 to 20 minutes prior -to anesthetic induction in non-emergency procedures. Sedative and anti-anxiety drugs include medications like acepromazine, chlorpromazine, diazepam and micfazelam. Some of these may be administered orally or injectably and serve to calm and relax dogs that may be anxious, hyperexcitable, in respiratory distress due to an airway obstruction, or frightened. Acepromazine is a tranquilizer that alone or combined with opioids provides excellent sedation. It has no analgesic properties, but has an excellent anti-anxiety effect. These drugs are used in much lower doses for minimal sedation than when they are used alone or in combination for anesthetic induction.

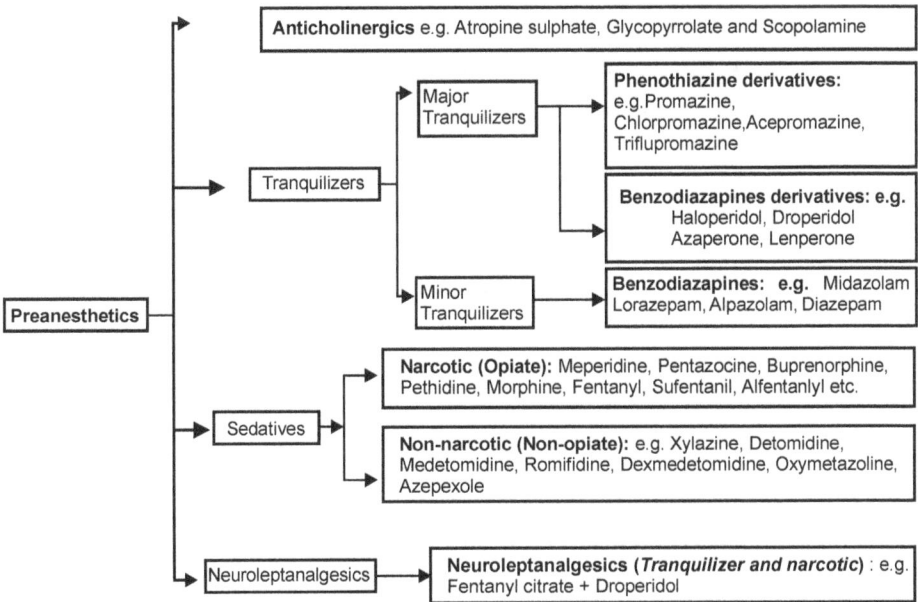

```
                        ┌─────────────────────────────────────────────────────────────┐
                     ──▶│ Anticholinergics e.g. Atropine sulphate, Glycopyrrolate and Scopolamine │
                        └─────────────────────────────────────────────────────────────┘
                                                          ┌──────────────────────────────┐
                                          ┌──────────┐ ──▶│ Phenothiazine derivatives:     │
                                       ──▶│ Major    │    │ e.g.Promazine,                 │
                                          │ Tranquilizers │ Chlorpromazine,Acepromazine,   │
                                          └──────────┘    │ Triflupromazine                │
                         ┌───────────┐                    └──────────────────────────────┘
                      ──▶│ Tranquilizers│                 ┌──────────────────────────────┐
                         └───────────┘                    │ Benzodiazapines derivatives: e.g. │
                                                          │   Haloperidol, Droperidol      │
                                          ┌──────────┐    │   Azaperone, Lenperone         │
                                       ──▶│ Minor    │    └──────────────────────────────┘
                                          │ Tranquilizers │─▶│ Benzodiazapines: e.g. Midazolam │
 ┌──────────────┐                         └──────────┘    │ Lorazepam, Alpazolam, Diazepam │
 │ Preanesthetics │───                                    └──────────────────────────────┘
 └──────────────┘                         ┌──────────────────────────────────────────────┐
                                       ──▶│ Narcotic (Opiate): Meperidine, Pentazocine, Buprenorphine, │
                         ┌──────────┐     │ Pethidine, Morphine, Fentanyl, Sufentanil, Alfentanlyl etc. │
                      ──▶│ Sedatives │─── └──────────────────────────────────────────────┘
                         └──────────┘     ┌──────────────────────────────────────────────┐
                                       ──▶│ Non-narcotic (Non-opiate): e.g. Xylazine, Detomidine, │
                                          │ Medetomidine, Romifidine, Dexmedetomidine, Oxymetazoline, │
                                          │ Azepexole                                     │
                                          └──────────────────────────────────────────────┘
                         ┌────────────────┐┌────────────────────────────────────────────┐
                      ──▶│ Neuroleptanalgesics │─│ Neuroleptanalgesics (Tranquilizer and narcotic) : e.g. │
                         └────────────────┘│ Fentanyl citrate + Droperidol              │
                                           └────────────────────────────────────────────┘
```

1. **Anticholinergics** - These agents are not used generally for imaging specially GIT and Cardiovascular system because they hamper the GI motility and produces tachycardia.

2. **Tranquilizers** - are useful in wide variety of conditions in animals, namely

    (*i*)   As preanesthetic sedative.

    (*ii*)   To relieve anxiety.

    (*iii*) To restrain refractory animals during examination or large animals during shipment.

- Tranquilizers are usually administrated intravenously at least 15 minutes prior to administration of general anesthesia.
- It can be given by intramuscular and oral routes of administration.

Phenothiazine tranquilizers

- Acetylpromazine is most commonly used tranquilizer in small animal practice.
- Phenothiazine tranquilizers can be given orally, subcutaneously, intramuscular or intravenous route.

Butyrophenone derivatives

Azaperone -It is a narcoleptic and probably the most potent and specific sedative available for swine.

Benzodiazapine tranquilizers: *Diazepam*

- It has calming, muscle relaxant and anticonvulsant activities in animals.
- It is an excellent preanesthetic agent for animals with a history of CNS disorders.
- Diazepam appears to act on parts of the limbic system, the thalamus, and the hypothalamus to produce calming effects or taming effect in animals.
- It can be used as preanesthetic in old and debilitated animals.
- In healthy animals, it may not produce CNS depression or tranquilization.
- Risk of congenital malforaaafibn during early pregnancy.
- Contraindicated in glaucoma.
- Flumazenil is highly potent antidote for benzodiazepines (0.25-1.50 mg/kg IV or IP.). Flumazenil has high affinity for benzodiazepine receptors and will reverse all the CNS effects of benzodiazepines.
- Its duration of action is about 60 minutes.

3. Sedatives

A. Narcotic analgesics

Morphine

- Morphine is usually injected subcutaneously or intramuscularly but may be given by slow intravenous infusion.
- Morphine is contraindicated for radiography of
    1. Uremic patients because it stimulates secretion of antidiuretic hormone (ADH) and urine production may reduce as much as 90%.

2. Patients with traumatic or hemorrhagic shock (due to severe depression of BP, cardiac output and lowered O_2 consumption).

Fentanyl citrate and Carfentanyl

- Carfentanyl is a congener of fentanyl which is 12000 to 18000 times more potent than morphine and has been termed *super fentanyl*
- Used almost exclusively for restraining of exotic species of animals.
- Action reversed by morphine antagonists, e.g. cyprenorphine or deprenorphine.

Etorphine HCl (M-99)

- It is an oripavine derivative and used extensively for restraint of wild animals.
- Etorphine is 80-1000 times more potent than morphine
- The effect of etorphine can be antagonized with
    (i) Nalorphine (etorphine /nalorphine ratio 1 : 1 0 to 1 :20)
    (ii) Diprenorphine (etorphine / diprenorphine ratio 1:1 to 1 :2).

B. Non opiate analgesics: (α_2-agonist)

They can produce analgesia, sedation, anticonvulsant and calming effect. Analgesia and sedation is produced by stimulation in the CNS.

Xylazine HCl (Rompum)

- It is a potent non-narcotic sedative and analgesic as well as muscle relaxant.
- Sedative and analgesic activities are related to CNS depression mediated by stimulation of α_2-receptors.
- Its sedative effect lasts for 1-2 hours and analgesic effect lasts for 15-30 minutes.
- Emesis (in normal doses) and acute abdominal distension (in large doses) occurs (Parasympatholytic effect → gastrointestinal atony → accumulation of gases. This feature makes radiograph interpretations of upper GI tact less certain.
- Ruminants regurgitate and become tympanitic under its effect.

Medetomidine

- It is 10 times more potent than xylazine and 6 times more potent than detomidine.
- Cardiopulmonary and respiratory effects are similar to xylazine.

- Atropine is more effective than Glycopyrrolate in preventing bradycardia caused by medetomidine.
- Antagonist: Same as xylazine. Atipamizole is the most effective because of its α_1 and α_2 selective ratio.
- It is recommended for young and healthy dogs.

Dexmedetomidine

- It is approved for use in cats as a sedative and analgesics for minor procedures. At the dose rate of 40 fig/kg IM, dexmedetomidine induces a moderate to deep levelof sedation and provides chemical restraint and analgesia sufficient for clinical examinations and procedures, including radiography and ultrasonography.
- Sedation and restraining can be achieved using dexmedetomidine in combination with butorphanol (0.2 mg/kg IM), buprenorphine (0.015 mg/kg IM), or diazepam (0.4mg/kg IV) in dogs undergoing hip radiographic examination and requiring hind limb manipulation (hip extended or stress radiographic views) for pelvic radiography. However, dexmedetomidine-butorphanol induces excellent sedation with sufficient muscle relaxation to allow for completion of diagnostic procedure.

α_2-antagonists

- **Yohimbine:** Cattle @ 0.125 mg/kg IV, Horse @ 0.075 mg/kg IV, Dog @ 0.1 mg/kg IV and Cat @ 0.5 mg/kg IV.
- **Atipamizole** is a specific antagonist for medetomidine.
- **Idazixan** @ 0.05-0.1 mg/kg body weight in all species.
- **Doxapram** @ 2 mg /kg IM is a general CNS stimulant.
- **Tolazoline** works better than yohimbine in cattle (0.44 mg/kg intravenously) It is also effective in sheep dog and cat @ 2.0 mg/kg b. wt.

4. Neuroleptanalgesics

These are the combination of neuroleptic (tranquilizer) and analgesic (narcotic) to enhance the CNS depressant effects of each drug.

Fentanyl citrate - Droperidol: (innovar -vet).

- Used to produce neuroleptanalgesia (it is a state of CNS depression and analgesia produced without the use of barbiturates and volatile anesthetics).
- Fentanyl citrate (0.4 mg) + Droperidol (20 mg).
- Dose in Dog - 1 ml/10-15 kg b. wt. I.V.
- This combination is sufficient for minor procedures like radiography.

- The mixture produces sedation, analgesia, immobilization, respiratory depression, panting and bradycardia.
- Advantage: Easy to administer, wide margin of safety, quite recovery and partial reversibility with Opioid antagonists.
- A mixture of 4-Aminopyridine (0.5 mg/kg) and naloxone hydrochloride (0.04 mg/kg) IV, antagonize the CNS effects of droperdol-fentanyl.

Etorphine - Acepromazine (Immobilon, L.A.)- *For large animals.*

- Used in horses, wild and zoo animals.
- It has been used to immobilize horses for minor surgery e.g. castration.

Etorphine - Methotrimeprazine (Immobilon, S.A.)- *For small animal*

Etorphine antagonist - *Deprenorphine (0.272 mg/kg IV).*

Table 3.1: *Dose rate of different preanesthetics in different animals*

S. No.	Preanesthetic agent	Canine (Dog)	Feline (Cat)	Ovine (Sheep)	Caprine (Goat)	Bovine (Cattle and Buffalo)	Equine (Horse)	Swine (Pig)
1.	Atropine sulphate	0.02-0.04 S.C./I.M	0.02-0.04 S.C./I.M	0.7	0.7	0.04-0.06	-	0.06-0.08
2.	Chlorpro-mazine	1.0 I.V. 2.0 I.M	1.0 I.M	1.0 IV	1.0 IV	0.3 IM		0.2-0.3 IM/IV max dose 100 mg
3.	Triflu promazine	1-2 IV, 2-4IM	4 I.M	1.0 IV	0.2-0.3 IM	0.1 IV max 40 mg	-	-
4.	Acepromazine maleate	0.03-0.1 Max 3 mg IV, IM or SC	0.03-0.1	-	0.5-1.0 IM	0.5-1.0 IM	0.02-0.05 IV IM, SC	0.5-1.0 IM
5.	Diazepam	0.5-1.01 IV, 1.0-	0.5-1.0 I.M	0.5-1.0 IM	0.1-0.3 IM	0.03-0.11M	0.05-0.2IV	2-4 IM, 1.0-2.0 IV
6.	Xylazine	1-2 IM	1-2 IM	1-2 IM	-	0.01 - 0.02 IM	1.0-2.1 IM, 0.4 - 1.1 IV	-
7.	Detomidine	5-20 µg/kg IM	3.0 µg/kg IM	-	-	-	10-40	-

INJECTABLE ANESTHETICS AND THEIR COMBINATIONS USED FOR RADIOGRAPHY

Anesthetic induction agents are designed to rapidly induce unconsciousness that is free from excitement and struggling, and which will allow the placement of an endotracheal tube, if necessary, for maintenance of general gas anesthesia. An "ideal" anesthetic induction or protocol for all situations does not exist. Induction drugs are chosen with the intent of providing the smoothest induction possible with the least amount of undesired effects.

Non-opioid intravenous anesthetics (such as thiopental sodium, ketamine, Telazol, diazepam and midazolam, propofol, and etomidate) produce sedation, amnesia and deep sleep. The other components of anesthesia (loss of sensation, reflex and motor function) often require the addition of opioid analgesics or inhalant anesthetics. A few of the more popular non-opiod IV anesthetics are-

Injectable Anesthetics are more commonly used for inducing anesthesia in a patient to avoid the struggling which can occur as the patient passes through the first two stages of anesthesia. Injectable agents have a faster effect when given in the correct dose. Injectable Anesthetic Agents:

  (I)   Barbiturates
  (II)   Dissociative Agents (Cyclohexamines)
  (III) Steroid Anesthetics
  (IV)   Imidazole Derivatives
  (V)   Alkyl phenols
  (VI)   Potent Synthetic Opioid Analgesics

(I) Barbiturates: Thiopental Sodium (Pentothal)

- It is yellow crystalline powder that can be dissolved in distill water or saline to make concentrations like 1.25%, 2.5, 5.0%, 10% before use. In small animals a 2.5% solution and in large animals a 5% solution of thiopental is used.

- It should be stored in a refrigerator at 5°-6°C to retard deterioration.

- One third of the calculated dose should be administered rapidly intravenously within 15 seconds to get over the initial excitement associated with second stage of anesthesia. Remainder is administered slowly "to effect". Surgical anesthesia is reached when the pedal reflex is abolished. Additional doses may be given to prolong anesthesia.

- Rapid intravenous injection is usually followed by slow irregular respiration and sometimes brief period of apnea *i.e.* induction apnea (Up to 1-2 min).

- Use of thiopental is contraindicated in the neonate and feline porphyra.

- Rapid injection technique is contraindicated in animals in shock or in those with uncompensated cardiovascular diseases.

- *Dose*

  Dog and cat

  - 20 mg/kg for surgical anesthesia (Used alone).
  - 10-15 mg/kg for surgical anesthesia (After premedication). Pig: 10-25 mg/kg with Azaperone.

(II) Dissociative agents (cyclohexamines): Ketamine HC1

- Although ketamine can be used by itself in small animals as a sedative for procedures which do not require muscle relaxation, for example radiography, it is commonly combined with other sedatives for synergistic effect and to counteract muscle rigidity.
- Ketamine is one of the most popularly used drugs in veterinary practice due to its margin of safety and compatibility with other anesthetics.
- Ketamine increases Woed pressure and intracranial pressure so it should not be used in patients with glaucoma^and head injuries.
- Physostigmine salicylate is a ketamine antidote. Mixture of 4-aminopyridine and can be used.
- Ketamine should be used in combination with or after premeditation. Dose$_v$ Dog and mg/kg IM, Horse 2.2 mg/kg IM

Other combinations of Ketamine used for general anesthesia: The major goals of

a. Xylazine-ketamine drug combination

- Xylazine-ketamine drug mixtures are among the most popular means of producing large animal anesthesia.
- Length of anesthesia depends on species, dose and route of administration.
- There is good analgesia, muscle relaxation and sedation.

b. Benzodiazepine/Ketamine Mixtures

- Gives a period of surgical anesthesia from five to ten minutes with a recovery time of 30 minutes to 1 hour.
- For minor procedures e.g. for radiographs, suturing wounds, biopsies, etc., anesthesia can be maintained with 'top-ups' of a third to half the induction dose.
- **Dose:** 0.5-1.0 mg/kg diazepam (or midazolam) + 5.0-10.0 mg ketamine IV.

(III) Steroid anesthetics: Saffan or Althesin

- It is used for inducing anesthesia in cats and maintenance for short procedures like radiography. The advantages of this drug are that it provides rapid smooth induction, good analgesia; good muscle relaxation and supplementary doses do not prolong recovery very much.

- Dose: Cat - 3-9 mg/kg IV; Pig - 6 mg/kg IV; Reptiles- 12-18 mg/kg IM.

(IV) Alkylphenol: Propofol

- It is a substituted phenol derivative (2, 6-diisopropyl phenol) which is a relatively new short acting intravenous anesthetic agent.
- It is marketed in sterile glass vials / ampoules and contains no preservatives. Since the vehicle is capable of supporting micromal growth and endotox in production. Any remaining content of the opened vial / ampoule must be discarded within 6-10 hours. It should not be kept overnight for use on next day.
- A single bolus dose of propofol provides approximately 10 minutes of anesthesia.
- Dog and cats completely recovered within 20-30 minutes.
- Its short duration of action and rapid recovery time is due to rapid redistribution from the brain to other tissues and organs as occurs with thiopental.
- Recovery is faster after propofol because the drug's rate of metabolism is 10 times more rapid than that of thiopental.
- It causes significant decrease in intraocular pressure and therefore it can be used in ophthalmic patients.
- It is suitable for patients with renal or hepatic impairment.
- Short procedures (radiography) can be performed after a single bolus injection.
- Dose: Dogs and Cat-
    - (i)   6-8 mg/kg without premedication.
    - ii)   2.2-4.4 mg/kg after premedication with sedatives, tranquilizers or opioid.
    - (ii) The calculated dose should be given slowly till effect and approximately one third of the dose is given every 30 seconds.

(V) Miscellaneous agents

Tricaine methane sulfonatepr

- Used as an anesthetic to immobilize amphibians, fish and other cold-blooded animals by complete bathing of small subjects, by gill spraying in large fish, or by injecting in larger species.

Etomidate

- Etomidate is used infrequently in veterinary medicine. It is a short-acting sedative-hypnotic that causes minimal cardiopulmonary depression, and as such, it is often an excellent choice of anesthetic induction drug in the compromised dog.

- Alpha-2-agonist drugs, xylazine and medetomidine, provide excellent sedation, analgesia and muscle relaxation. Dexmedetomidine is more potent and provides a longer duration of analgesia; however it should be used only in young, healthy animals. Its advantage is that it is reversible and may administer as an intravenous or intramuscular injection.

Anesthetic maintenance

If our dogs are undergoing general anesthesia, they are usually maintained on inhalant, gas anesthetics, unless a constant rate infusion of an injectable anesthetic, like propofol above, is used. The two most common inhalants used today are isoflurane and sevoflurane.

INHALATION ANAESTHETICS USED FOR RADIOGRAPH

Methoxyflurane:

- Maintenance of anesthesia can be accomplished with concentration of 0.4-1.0%. It is highly rubber soluble. Both induction and recovery may be prolonged due to uptake and release of methoxyflurane from rubber in the anesthetic circuit.
- It causes a dose dependent depression of the cardiopulmonary system. The heart rate and blood pressure generally remains stable due to the release of endogenous catecholamine.
- It produces good analgesia and muscle relaxation.
- It produces renally toxic fluoride ions so should not be used with renally excreted drugs such as tetracycline and flunixin.

Enflurane

- Induction and recovery are rapid. Maintenance concentration should not exceed 3%. At high concentrations (> 3.5%) there is increased central stimulation and seizures like activity on the electroencephalogram. So it should be avoided in seizure prone patients.

Halothane

- Halothane in concentration of 2-4% produces stage III, plan 2-3 anesthesia in dogs. Surgical anesthesia is maintained by inhalation of 0.8-1.2% mixtures.
- There is dose dependent depression of the cardiopulmonary system with halothane and the blood pressure is lowered by direct myocardial depression as well as vasodilatation through depression of the vasomotor tone center of the medulla.

- Halothane sensitizes the myocardial conduction system to the action of epinephrine and nor-epinephrine. Arrhythmias may result.
- Halothane is safe anesthetic during severe anemia.

Isoflurane

- It is less potent than halothane or methoxyflurane and relatively insoluble leading to fast induction and recoveries.
- The MAC for isoflurane does not change with duration of anesthesia. MAC for the dog is 1.28% and for cats is 1.63%.
- It can initiate fatal hyperthermia in susceptible swine.
- It does not promote seizure activity.
- Due to its good cardiac stability, the use is indicated in patients with some cardiac dysfunctions. This is especially true for patients with arrhythmias which may worsen with halothane.

Desflurane

- Desflurane is totally fluorinated ether. It needs new vaporizer technology which has increased the cost of using desflurane.
- Its use in veterinary anesthesia is limited by economics.
- Circulatory and respiratory effects are similar to isoflurane.
- It is the least soluble of the volatile inhalation anesthetic agent in blood and body tissues MAC in dogs is 7.2%.

Sevoflurane

- It should be used only in precision, agent specific, out of circuit vaporizers.
- CNS and cardiovascular effects are similar to isoflurane and desflurane.
- Dose dependent decrease in cardiac output and blood pressure.
- It does not sensitize the myocardium to catecholamines.
- Dose dependent CNS depression.
- MAC in dogs is 2.36 % and in cats is 2.58%.

RADIOGRAPHY OF DIFFERENT BODY SYSTEMS UNDER SEDATION/ANESTHESIA

Diagnostic radiographs are taken to evaluate orthopedic injuries or pain, to evaluate the lungs and heart in states of disease or distress, in an emergency situation for triage, or to evaluate other potential abnormalities of other body systems. Effective radiology techniques are designed to 1) obtain high-quality, diagnostic radiographs, 2) limit exposure of the patient and staff to excessive X-ray beam exposure, 3) be as pain-free and stress-free as possible for the patient, especially in a critical or painful animals. Emergency radiographs are taken in an emergency situation in order to provide the information to the veterinarian for diagnosis as well as in formulating an emergency therapeutic plan. Positioning is often determined by the injury to the dog, the patient's overall stability, and the purpose for the radiograph. Often no sedation to minimal sedation is used for radiographs in cases of bloat (GDV), vehicular trauma, difficulty whelping, fresh fractures or gunshots.

Orthopedic Radiography

The orthopedic ailments like fractures, sprains, strains or joint swelling are extremely painful to animals and require early diagnosis and treatment. To obtain a diagnostic-quality radiograph, especially when evaluating for pre-surgical films, or for submission to a radiologist for evaluation, sedation/analgesia is an essential and valuable tool. Positioning of the affected joint or limb is crucial to evaluate bones and soft tissues for abnormalities, especially when dealing with areas where subtle changes may exist. Radiographs in these cases are not only used to tell if an abnormality exists, but often to plan a surgical repair. Good positioning is achieved when a patient is relaxed and still, and when a veterinary staff is able to use techniques such as foam padding, soft-tape, positioning-boxes and other "tools." We cannot ask our animals to "hold still" and as such, we must position them. If they are in pain, or if a joint is swollen, positioning for-a diagnosis may certainly be uncomfortable. Analgesics are helpful in providing pain-relief and in relaxing the patient for these procedures.

Sedation and analgesia although not required but sometimes sedation is recommended for orthopedic radiographic study as it allows the patient to relax, thereby allowing easier positioning (particularly extension for cranial caudal images). In addition sedation also decreases the amount of time required for repositioning due to lack of patient cooperation. For animals with painful joints, a neuroleptanalgesics protocol provides both sedation and pain control.

The animals should be approached appropriately and every effort should be made to calmly interact with them. Struggling with a frightened or painful animal, or forcefully holding them down is stressful to the patient and staff, will take longer to take the radiographs, and achieves nothing. Many times, poor radiographs will have to be repeated, causing repeated stress to the patient and costing more for additional films.

Radiography of the Hip Joint

For diagnosis of hip dysplasia, the animal is restrained in dorsal recumbency with extended hind limbs and rotated inward. This position is uncomfortable, but crucial if we are going to get a good quality radiograph and see the (sometimes very small) abnormalities which may be present. Cats may be given small amounts of gas anesthesia through a mask for several minutes while the radiographs are being taken. In canines sedatives/ anesthetic drugs is given by injection, and 10-20 minutes later the dog has responded to the light anesthesia. After taking the radiograph, give a second injection of specific antidote to reverse the sedation. In either case, the dog or cat is usually ready to walk out to the waiting room 15 minutes later. And we have high quality radiographs.

The OFA radiographic view and PennHIP radiographic views are specifically taken for evaluating a quality radiograph of hip joint laxity and/or the likelihood of the joint to luxate or subluxate in canines. Lack of sedation or anesthesia may affect a regular OFA view in that we may be falsely making the hip joint "look better" via the surrounding musculature being tensed. etc. Conversely, you cannot make a "good hip" look bad with anesthesia. For the best and most accurate assessment, the dog should be under general anesthesia.

Sedation of animal is important to achieve a better quality roentgenogram and thus helps to better diagnose subtle changes that may be present radiographically in regard to evidence of degenerative joint disease (DID) in the joint on an OF A view. Taking a near-perfect OF A radiograph is a challenge. With an anesthetized dog in a positioning device (such as a foam-padded V-tray), it is easy to make positioning corrections and approach that perfect view. With an awake and/or restrained dog, you are forced to restart the process each time you make an exposure. Errors in positioning can either overstate or underestimate the true status of the hip joint. Also, to obtain diagnostic quality radiographs; the musculature around the hip joint must be completely relaxed. For the comfort and safety of the animal, this requires either heavy sedation or general anesthesia.

Dexmedetomidine along with buprenorphine, butorphanol or diazepam have been used for restraining and sedation of dogs during radiographic examination of hip (extended or stress-liography views). Dexmedetomidine and its combination with butorphanol or diazepam produces excellent sedation in dogs. However, combination involving buprenorphine had overall a relatively poorer quality of sedation and required additional administration of buprenorphine before the radiographic procedure could commence. Once sedated, clinically sufficient muscle relaxation accompanied by a very low proportion of dogs responding to pain stimuli were observed. Dexmedetomidine sedative protocols, particularly in combination with butorphanol and diazepam, can be used effectively and safely in dogs for radiographic procedures.

Radioglaphic appearance of the coxofemoral joints was evaluated by taking pelvic radiographs of dogs with or without anesthesia. There was no statistical difference between the two groups of dogs. This study failed to demonstrate any changes due to anesthesia on the radiographic appearance of the coxofemoral joints. Anesthesia may, however, be beneficial for proper positioning and to decrease unnecessary patient and personnel exposure to radiation.

Vertebral Fractures and Luxations

Anesthesia is a prerequisite for accurate positioning for radiography of the spine as spinal conditions are painful and the necessary positioning will not otherwise be tolerated.

For radiographic examination of vertebral column, the dogs can be sedated with medetomidine 10-20 µg/kg intramuscularly (IM) and methadone 0.1-0.2 mg/kg IM.

When sedation is adequate for optimal radiographic imaging, anesthesia is not induced before performing the radiography or MRI. If sedation is inadequate for perpendicular positioning of the vertebral column to the film, anesthesia may be induced with propofol and maintained on isoflurane and O_2. A good quality lateral and ventro-dorsal survey radiographs of the entire vertebral column, from Cl to S3 can be taken by using these sedatives/anesthetics.

After securing the patient in lateral recumbency and doing the initial physical and neurological examination, the animal can be sedated for lateral spinal radiographs, preferably with an opioid to provide pain relief. It is important that the radiographs be as straight as possible without moving or struggling with the patient. Rotated radiographs can be extremely difficult to interpret, particularly in the cervical region. In some cases, heavy sedation or anesthesia is required for proper positioning. A lateral thoracic radiograph should be taken at the same time to look for pulmonary trauma. Some fractures/luxations do not cause significant displacement, and radiographic findings may be subtle. Remember that radiographs are-static images, and it is possible for luxation to occur followed by realignment of the vertebral column, making radiographs appear normal when there is actually vertebral instability. Radiographs should be taken carefully because heavy sedation and anesthesia relax paraspinal musculature and can further destabilize the spine if held together by muscle contraction.

Myelography

The spinal cord is not visible in the conventional roentgenogram. Therefore different contrast procedures have been developed to visualize the spinal cord (myelography), the epidural space (epidurography) and the intervertebral disk (discography). The procedures are performed on the patient in narcosis. After the injection of contrast agent at the site, the normal radiograph is taken and the image evaluated. The myelography is frequently quicker than MRI and so the duration of narcosis can be shortened.

Phenothiazines should not be used as preanesthetic sedative/tranquilizer for myelography of patients suffering from seizures because it lowers the seizure threshold and aggravates the condition.

Oral Radiography

Sedation or anesthesia is necessary for dental radiography so that the animal can be properly positioned. Dental radiographs are generally obtained during a routine dental examination and cleaning. However, in cases of facial trauma or head trauma, dental radiographs may be taken to assess the extent of damage to the mouth, teeth, and jaws. Sedation is needed so that the animal can be properly positioned for the radiographs to be taken.

The animals are mostly presented to the hospital with full stomach for radiography. In emergency cases, radiograph should be taken immediately. Otherwise it must be postponed and the owner should be advised to off feed the animal for 12 hour before taking the radiograph. So that the animal can be sedated or anesthetized to take the best quality radiograph to diagnose their condition because most pet animals are anxious and could not be restrained in correct position. Oral radiography is indicated in small animal practice for the diagnosis and treatment of persistent deciduous teeth, periodontal disease and fractured roots during extractions, tooth resorption, endodontic disease with or without pulpal exposure, maxillary or mandibular fractures, and oral neoplasia.

Thoracic and Abdominal Radiography

Thoracic or abdominal radiographs are taken to evaluate for evidence of heart or lung disease, the presence of metastatic cancer, gastrointestinal abnormalities, urinary tract disease or other problems. Positioning for evaluation in these cases is often just as important as in orthopedic films. On radiographs of the thorax, films taken during inspiration versus expiration may look dramatically different, and motion during the "shot" will interfere with interpretation. Mild sedation is helpful to relax the patient, and in some cases, such as pancreatitis, analgesics may help a painful pet to be more comfortable.

Thoracic Radiography

For radiography of the thoracic cavity, sedation is indicated to avoid rapid respiratory movements or panting. General anesthesia permits manual inflation of the lungs and maximises aeration. The advantages for this are:

- The prevention of artifacts associated with increased lung opacity due to poor inflation (often seen in conscious or sedated patients).
- Increased air within the lungs reduces the lung tissue density.
- It allows holding the breath for a long as necessary to make the exposure. This minimizes the movement blur that is likely when using the prolonged exposure times needed with a low output X-ray unit.

There are also several disadvantages with general anesthesia:

- Performing general anesthesia depends on the condition of the animal as described by ASA.
- Prolonged lateral recumbency leads to partial collapse of the dependent lung and subsequent hypostatic congestion which is more marked with general anesthesia. It produces increased lung opacity of the dependent lung which can lead to a wrong diagnosis. While the animal is in lateral recumbency, the heart will drop towards the dependent chest wall making some of the peripheries of the ventral lung field retract from the chest wall. This should not be mistaken for a pleural effusion. To avoid these unwanted opacities on the radiograph, it is best to take the dorsoventral or ventrodorsal projections before the lateral recumbent views. During any period of inactivity it is recommended to keep the patient in sternal recumbency to allow the lungs to re-expand.

- Manual inflation should be performed gently to avoid tracheal or lung tissue tear/rupture. For this reason, and to avoid over inflation, it is advisable to use un-cuffed endotracheal tubes in cats.

Cardiac Radiography

Many veterinarians are apprehensive about sedating an animal with a potential heart problem, particularly one that is dyspneic. Sedate these patients first, *i.e.* before radiographs. Choice for sedating cardiac patients, both dogs and cats, is diazepam and butorphanol, 0.2 mg/kg of each, mixed for an IV injection. Substitute midazolam for the diazepam (0.2 mg/kg again) and give it IM for the patient that can't even sit still for an IV. Butorphanol can be substituted by oxymorphone (0.05 - 0.10 mg/kg) or hydromorphone (0.10 - 0.20 mg.kg) but use the lower end dosage of the narcotic for cats.

Other sedative choices recommend against for cardiology patients include acepromazine, atropine or glycopyrrolate, ketamine or tiletamine (in Telazol) for cats with cardiomyopathy and particularly alpha agonists such as xylazine or medetomadine. BAG (butorphanol, acepromazine, and glycopyrrolate) increases the dog's heart rate over 300, the echocardiogram was uninterpretable". Diazepam and butorphanol combination can be used for taking an X-ray, doing an echocardiogram, or performing a pericardiocentesis, thoracocentesis, or abdominocentesis.

Abdominal Radiography

Abdominal radiography is useful for evaluating the size, shape, and position of abdominal organs. Sedation is sometimes recommended for patients undergoing abdominal radiography. When an animal is sedated, positioning is much easier because the patient is completely relaxed. Sedation may also be recommended if the patient is in pain. The imaging procedure is safe and completely painless and can be performed on calm and cooperative pets without sedation. The veterinarian may administer a sedative or general anesthesia in cases where a dog or cat has trouble becoming fully relaxed naturally. Patients are lightly sedated for radiology and in some cases, they may require general anesthesia.

Barium Meal

A barium meal includes food mixed with barium sulphate being administered to a patient for observing its passage through the gastrointestinal tract. The food should be with held for a minimum of 24 hours. A barium concentration of 80-100% W/W can be administered into the cheek pouch of the patient using a 50mL syringe with a catheter tip, and allow the patient to swallow it in its own time. Administration of the contrast material in the'fractious animals, a tube may be used to administer the barium.

- For barium meal, acetylpromazine is the best sedative agent in dogs as it doesn't significantly alter gastrointestinal function.
- The use of compounds such as atropine, ketamine, and barbiturates significantly depress gastrointestinal motility, and are not recommended in dogs.
- Ketamine/Diazepam is the choice of preanesthetic in cats for barium meal. (In cats a combination of ketamine (3mg/kg) and diazepam (0.1 mg/kg) is an effective sedative which doesn't affect the gastric transit time.

Excretory Urography

Some procedures, such as radiography of the bladder that need contrast material to be put in through a urinary catheter or hip score X-rays that need the dog to be positioned exactly on its back, requires a full anesthetic. The animal should be prepared for radiography. For this the patient is off feed for 36 hours prior to radiography and ensures evacuation of the large intestine prior to commencement. The use of positioning aids assists in ensuring adequate positioning for each radiograph. While the procedure can be done on a conscious or sedated patient, anesthesia eliminates personnel exposure while enabling rapid and precise repositioning during the procedure.

REFERENCES

Ian J. Flaws (2010). The evolution of oral radiography in veterinary medicine. *Can Vet J.*, 51(8): 899-901.

Everett Aronson, Karl Fl. Kraus and Julif Smith (1991). The effect of anesthesia on the radiographic appearance of the coxofemoral joints. *Veterinary Radiology*, 32(1): 2-5.

Cecilia Rohdin, Janis Jeserevic, Ranno Viitmaa and Sigitas Cizinauskas (2010). Prevalence of radiographic detectable intervertebral disc calcifications in Dachshunds surgically treated for disc extrusion. *Acta Veterinaria Scandinavica*, 52:24.

Erden IA, Pamuk AG, Akinci SB, Koseoglu A, Aypar U. (2010). Comparison of two ketamine-propofol dosing regimens for sedation during interventional radiology procedures. *MinervaAnestesiol.* 76(4):260-5.

Leppanen MK, McKusick BC, Granholm MM, Westerholm FC, Tulamo R, Short CE. (2006). Clinical efficacy and safety of dexmedetomidine and buprenorphiner butorphanol or diazepam for canine hip radiography. *J Small Anim Pract.* 47(11): 663-9.

Steyn P.F. and Twedt D.C. (1994) *J. Am. Anim. Hosp. Assoc.* 30:78-80.

Leppanen, M.K., Mckusick, B.C., Granholm, M.M. et al. (2008). Clinical efficacy and safety of dexmedetomidine and buprenorphine, butorphanol or diazepam for canine hip radiography. *J. Small Animal Pract,* 47 **(11):** 663-669.

4 Production and Properties of X-rays, Factors Influencing Production of X-rays

X-rays are one of the electromagnetic radiation and widely used for diagnosis and treatment.

PROPERTIES OF X-RAYS

1. Wave length of X-rays is 0.05-0.01nm (shorter than visible light). This is the reason that X-rays are able to penetrate objects.
2. X-rays ionizes atoms and molecules of a substance.
3. X-rays expose the photographic emulsion and produces latent image on photographic film which is converted in to the visible image by processing of exposed X-ray film.
4. X-rays causes fluorescence by certain substances.
5. It can ionize the gases.
6. It has no charge and mass.
7. Travel at the speed of light i.e. 3×10^8 m/s
8. X-rays are invisible and travel in a straight line.
9. They cannot be deflected by magnetic field.
10. They penetrate all matter.

X-RAY TUBE

X-rays are produced within a vacuum X-ray tube when fast moving electrons collide with matter. X-ray tube is thermionic diode which includes:

1. Cathode
2. Anode
3. Glass envelope
4. Beryllium window

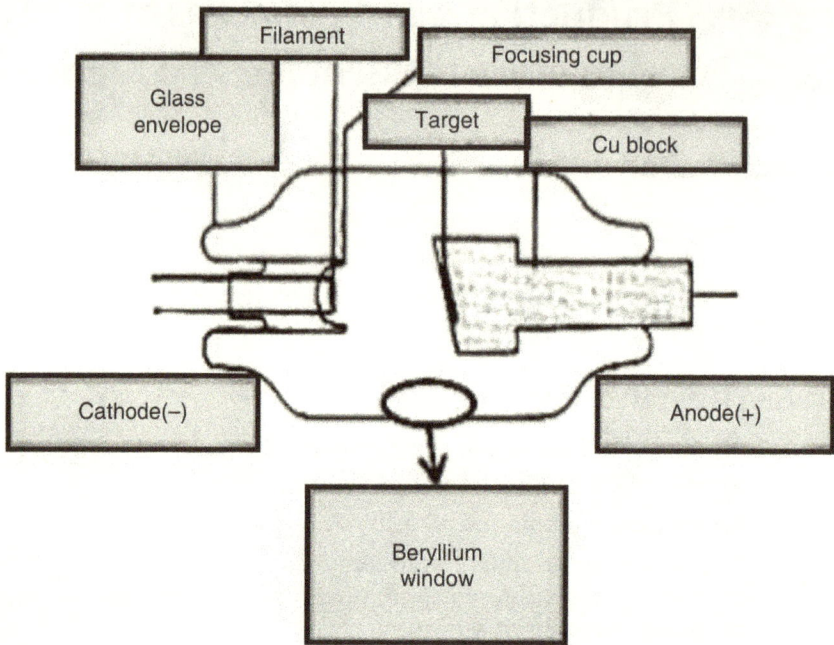

Fig. 4.1: X-ray tube

1. **Cathode**
   - It is a coiled tungsten wire (0.2 mm diameter) which is housed within a focusing cup.
   - **Focusing cup:** The cathode filament is embedded in a concave metal (Molybdenum or Nickel) shroud known as focusing cup. It focuses the beam of electrons on the focal spot present on the anode target.
   - When the current is passed to the filament, the filament is heated to supply a cloud of electrons and more X –rays are produced.
   - The milliamperage (mA) controls the amount of radiation production.
   - The newer machines are advantageous due to-
     (i) The cathode is made up of tungsten rhenium alloy which prolongs the cathode life along with enhanced thermionic emission efficiency.
     (ii) Double filament is present which is mounted side by side or one above the other. Generally one filament is larger than the other and only one filament is used at a time. The larger filament is used for larger exposure.

Why tungsten is chosen for X-ray tube-
   1. High melting point (3370^0C) to withstand higher tube current.

2. High atomic number (74) provides higher electron emission.

3. Fewer tendencies to vaporize to prolong the life of the tube.

2. **Anode**

    *   It is also made up of tungsten which forms the positive terminal of X-ray tube.

    *   Anode is the actual site of X-ray production.

    *   **Focal spot:** The focal spot is the area of anode (target) which is bombarded by electrons from the cathode. The focal spot is oriented at 11^0 to 20^0 angles.

    *   **Effective focal spot:** It is the actual focal spot.

        Advantage: Small effective focal spot produces better detail.

        Disadvantage: Less dissipation of heat.

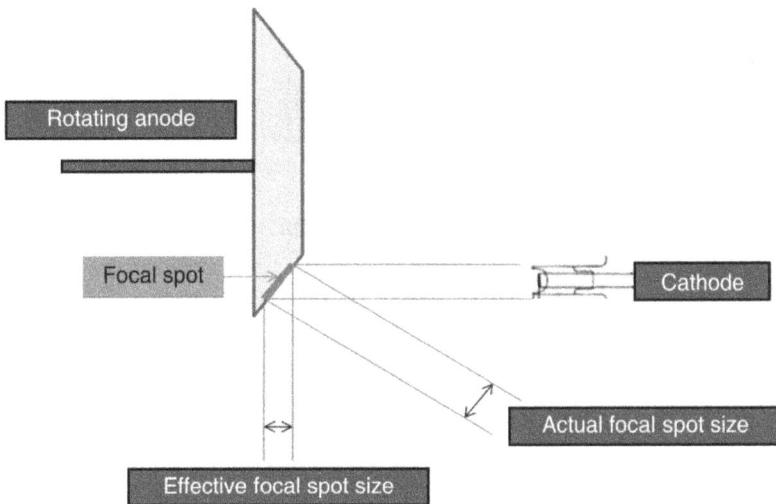

Fig. 4.2: Line focus principle

    *   *Effective focal spot size principle:* By angling the anode, the size of the effective focal spot is decreased whereas the area on the anode that is stricken by electrons remains large and thus facilitates heat dissipation. The effective focal spot is smaller than that of the actual focal spot.

    *   Some machines are equipped with both small and large focal spots. A small focal spot have shorter filament and focusing cup as compared to large focal spot.

    *   Anode are of two types-

        (*i*) *Stationary anode:* used in low output machines with low mA capabilities e.g. (5-50 mA). In stationary anode a block of tungsten is embedded into a copper anode.

> **Why copper is used: Because copper is a good conductor of heat which helps in dissipation of heat**

Disadvantage: After some time the target becomes pitted because electrons continually struck at the same target region.

(*ii*)   ***Rotatory anode:*** Used in higher output machines with high capabilities (100-1200 mA). It is a disc of molybdenum alloy with covering of tungsten at the periphery. Rotating anode tubes may contain a focal spot of about $1/6^{th}$ the size required for stationary anode tubes. The anode rotates at high speed during exposure. The friction of rotating anode is reduced by self lubricated bearings coated with metallic barium or silver.

Advantage: Dissipation of heat over more surface area while maintaining a relatively small effective focal spot. It prevents the pitting of target.

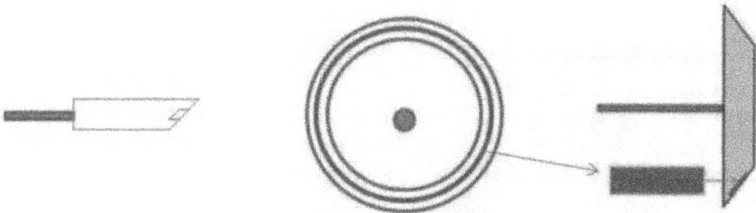

Fig. 4.3a: Stationary anode   **Fig. 4.3b:** Rotatory anode

3.   **Glass envelope:** The cathode and anode are housed in a vacuum glass envelope made of pyrex or borosilicate glass. Vacuum allows unobstructed path for electron beam/tube current and prevents oxidation and burning out of the filament. The glass envelope is housed in a metal covering which is lined with lead except at window. The protective metal housing contains sealed oil which serves as an electrical insulator and helps in heat dissipation.

4.   **Beryllium window:** The X-ray tube has thin glass or beryllium window which allows maximum emission of X-rays with minimal absorption in the glass envelope.

DIFFERENCE BETWEEN DIAGNOSTIC AND THERAPEUTIC X-RAY TUBE

S.No.	Diagnostic X-ray tube	Therapeutic X-ray tube
1.	Operated at 10-1200mA, 0.001-10 seconds, 20-150 KVp	5-20 mA, relatively long period of time, high KVp setting e.g. • Low voltage (50-120 kV) for treatment of cutaneous ailments. • Intermediate voltage (130-150 kV) for treatment of lesions located a few cm beneath the skin.

Contd...

Contd...

		• Orthovoltage (160-300 kV) for treatment of deep seated lesions. • Mega or supervoltage (300-1000 kV) for treatment of very deep seated lesions.
2.	The cathode or filament may be single or double	Single large filament
3.	The anode is mostly rotating with small focal spot size.	Stationary with much larger focal spot size (5-8mm). Large tungston target and massive copper anode as compared to diagnostic X-ray tube.
4.	Target angle is < 300	>300
5.	Cooling of tube by sealed oil	Cooling equipment is much heavy with more amount of oil. Cooling is also by circulation of cool water through a coil within the shield

TRANSFORMERS

High voltage potential is required to accelerate the electrons from the cathode to the anode.

1. ***Step up transformer:*** The high voltage required on the anode side which is produced by step up transformer

Incoming voltage	Output voltage
1. In small portable machines: 110- 120 volts. 2. In large machines: 440-880 volts	40-150 kilovolts depending on machines

2. ***Step down transformer:*** Relatively low voltage is required by the filament on the cathode side (filament).

   • Incoming voltage is decreased and amperage is increased to 5-1200 mA.

   • Increase in amperage in the filament increases the number of electrons produced by the filament and thus production of number of X-rays is increased.

   • Generation of total no. of X-rays is directly proportional to the mA.

   • KVp controls the quality of X-ray beam.

   • mA and exposure time affects the quantity of X-rays produced.

RECTIFIERS

While using alternating current (A.C.) free electrons on the surface of the anode may be attracted towards the cathode and may damage it. To overcome this problem, A.C. is converted to D.C. by rectification. The rectifier limits the electron flow to one direction i.e. from cathode to anode. X-ray machines are rectified in 3 ways-

S. No.	Comment	Half wave	Full wave	Three phase
1.	Current	Single phase	Single phase	Three phase
2.	Flow of electrons from cathode to anode	Only on the positive phase of incoming alternating current	Positive phase is used and negative phase is also converted to a positive phase	Constant flow of electrons
3.	Efficiency	Less efficient	Efficient	Most efficient
4.	Disadvantage	Only positive phase of the cycle is used for X-ray production	-	-
5.	Pulse/second	60	120	-
6.	Exposure time	Longer exposure time is required	Faster exposure time	-
7.	Electron flow or X-ray production	**+potential** (electron flow or X-ray production) **-Potential** (No electron flow or No X-ray production)		

TUBE RATING CHART

It is necessary to run the X-ray machine within safe exposure limits to prolong the life of X-ray tube. The exposure characteristics are-

(i)   mA

(ii)   kVp (Kilovoltage peak)

(iii) Exposure time and

(iv)   Focal spot size

- Tube rating chart is different for different exposure characteristic.
- Wrong exposure selection may damage the tube e.g.
    1. *Tube overload:* When kVp and mAs are too high, more heat is generated causing cracking of anode.
    2. *Tube saturation:* At too low kVp the positive potential between the cathode and anode is not sufficient to pull all the electrons across the tube. It may cause tube cracking and electroplating (extra electrons build up on the glass envelope). Setting of mA below 500 with 60kVp or less should be used to check the tube saturation.

HEEL EFFECT OR ANODE HEEL EFFECT

Emission of unequal X-ray beam intensity from the X-ray tube is known as heel effect. The distribution of X-ray beam intensity decreases rapidly on the anode side of the tube. The X-rays are absorbed by target and anode material resulting in lower intensity towards the anode side. The heel effect is advantageous in the radiographic area of unequal thickness. Place the thicker area towards the cathode side to provide the more even film density.

Heel effect is more pronounced when:

1. Large X-ray films are used.
2. Short focal film distance is used.
3. lower target angle tubes are used

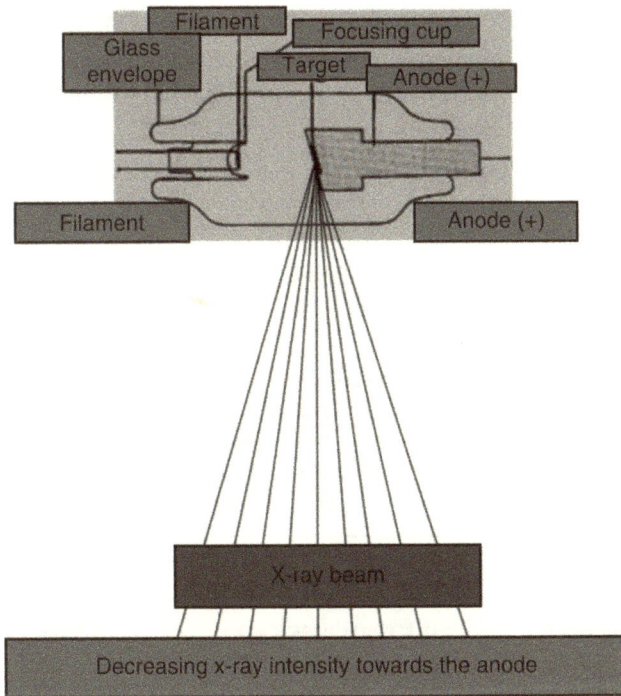

Fig. 4.3: Heel effect

PRODUCTION OF X-RAYS

- Electric current (mA) is passed through the filament.
- The heat produced by electric current allows electrons to boil off the surface of the filament and forms electron cloud around the filament. These electrons are stationary.

- When high potential voltage difference (kVp) between cathode and anode is applied, electrons are accelerated towards the anode. This stream of electrons is called as tube current.
- When electrons strike the anode, X-rays are produced by either collisional or radiative interaction.

1. **Collisional interaction**
   - It produces characteristic X-rays.
   - It involves a collision of a high speed electron from the cathode and an atom in the anode.
   - The high speed electron ejects an electron from the K shell of an atom in the anode.
   - The outer shell electron fills the void in the inner shell.
   - The difference in the binding energy is emitted as an X-rays photon.
   - X-rays created by collisional interactions accounts for only a small fraction of the total X-rays produced in a diagnostic X-ray tube.

2. **Radiative interaction**
   - The high speed electron passes close to the nucleus of the target atom (attracted by the positive charge) but electron is not ejected from the atom.
   - The electron slows as it binds towards the nucleus and releases energy in the form of electromagnetic radiation called *Bremsstrahlung or braking radiation*.
   - The energy released depends on the energy lost from the electrons during deflection by the nucleus.

> **Most of the energy used in production of X-rays is converted in to heat (99%). Only 1% is utilized for X-ray production**

Component of X-ray generator

1. High voltage circuit: regulates kVp
2. Filament circuit: Regulates mA
3. Circuit of the automatic exposure control panel.

Interaction of radiation with matter: Radiographs are produced by interaction of X-ray photon with matter by the following mechanism:

1. Coherent scattering ⎫
2. Photoelectric effect ⎬ Important in diagnostic radiology
3. Compton scattering ⎭
4. Pair production
5. Photodisintegration

1. **Coherent scattering**
    - X-ray photon interacts with object and deflects without changing its energy is called coherent scattering.
    - Coherent scattering (about 5%) degrades the quality of image (film fog) and increases exposure to the persons due scattering of X-ray photon.

2. **Photoelectric effect**
    - Interaction of X-ray photon to the tissue atom ejects the orbital electron (photoelectron) from the inner shell of tissue atom and energy of the photon is completely absorbed (**Fig. 4.4**).
    - Photoelectron produces more tissue ionization and eventually absorbed in the patient.
    - Probability of photoelectric interaction is directly proportional to the cube of photon energy $(1/E^3)$ that's why photoelectric absorption result in more difference in absorption and good contrast between tissues.
    - Advantage: Lack of scattered photon.
    - Disadvantage: Patient dose is high.

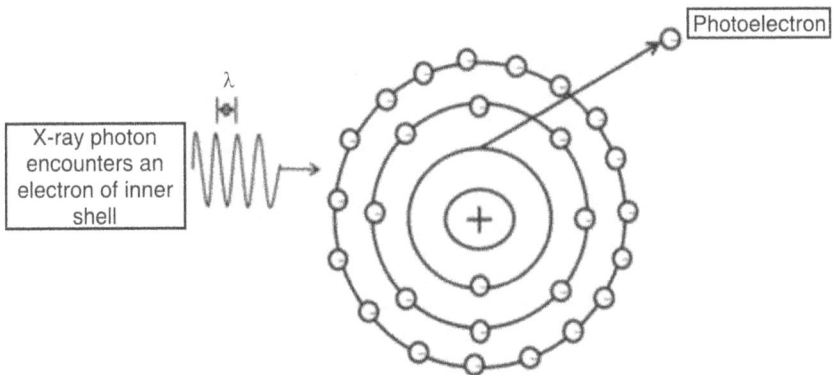

Fig. 4.4: Interaction of X-ray photon to the tissue atom ejects the orbital electron (photoelectron) from the inner shell of tissue atom and energy of the photon is completely absorbed.

3. **Compton scattering:** Scattered radiation in the diagnostic radiology is mostly due to compton scattering. In compton scattering process-
    - An incoming X-ray photon interacts with the peripheral shell electron of tissue atom.
    - The electron is ejected and photon is scattered at an angle **(Fig. 4.5).**
    - The scattered photon is also called as Compton electron or a recoil electron which has low energy than the original photon.
    - Compton reaction depends on total number of electrons in the patient.

- The probability of the Compton interaction decreases as the photon energy increases.
- Compton reaction is independent of atomic number that's why it results in poor image contrast.
- Scattered photon degrades the image by producing film fog.

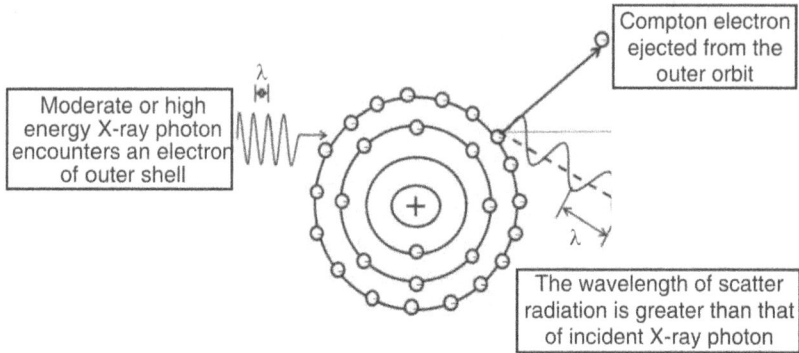

Fig.4.5: Compton scattering

TUBE EFFICIENCY FOR X-RAY PRODUCTION

Tube efficiency can be defined as thepercet of kinetic energy of projectile electrons that is converted to X-rays. It is directly proportional to the atomic number of the target, KVp and tube current.

$$\text{Tube percent efficiency (E)} = \frac{\text{X-ray power output}}{\text{Cathode rays power (watt)}} \times 100$$

$$\text{Tube percent efficiency (E)} = \frac{KZIV^2}{IV} \times 100$$

$$E = KZV \times 100$$

Where, $K = 1.4 \times 10^{-7}$ Volts-1

Z-Atomic number of target material

V-RMS value of tube voltage (KVp)

I-RMS value of tube current in amperes.

X-ray Tube Rating Chart

Every X-ray tube has an individual tube rating. X-ray tube ratings dictate the maximum combinations of kilovoltage peak kVp, milliamperes (mA) and time that can safely be used without overloading the tube. Rating is expressed in kilowatts. Both electrical and thermal limitations exist for a given X-ray tube.

- kVp is set by insulation consideration.
- mA is limited by permissible filament temperature.

Selection of chart depends on-

- **Power supply:** Single phase or three phase. Both have different tube rating.
- **Voltage rectification:** Maximum operation limit is more with full wave rectification.
- **Focal spot:** Maximum operation limit of machine is less with a small focal spot.
- **Application:** In serial angiography and tomography, a special chart is used.

1. **Radiographic tube rating chart**
   - It indicates maximum safe exposure limit for selected kVp and mA for a single exposure.

2. **Anode heat cooling chart**
   - Anode heating capacity is measured in heat units which is the product of kVp x mA x Exposure time.

     1 heat unit= 0.785 joule

   - Indicates maximum heat units that may be safely stored in the anode and also the time required for anode cooling between the exposures.

3. **Tube housing cooling chart**
   - Tube life depends on ability of metal tube housing to store and dissipate heat.
   - The temperature of tube should not exceed above 90°C.

6

Exposure Variables

Radiographic density, contrast and detail are controlled by exposure variables like mAs, kVp, FFD and OFD. These variables depend on the thickness and nature of the part to be radiographed.

- Thicker parts need more exposure than thinner ones.
- Nature of the part: Depends on tissue density like-
  - Hard tissues like bone: increase kVp 5-10.
  - Soft tissues: For heavy muscled area- double mAs
    For less muscular area like cervical region- Decrease kVp 5-10
  - Thorax contains air- Half mAs
  - If contrast agent is used increase kVp 5-10 and mAs should be doubled.
  - If the part is covered with plaster cast increase the mAs (doubled).

1. Millamperage and exposure time (mAs)

- mA controls the number of electrons generated at the filament.
- Increasing the number of mAs increases the radiographic density by increasing the number of X-rays generated e.g. mAs is doubled-radiographic density is doubled.
- Total number of X-rays generated is controlled by exposure time.
- mAs is the product of milliamperage and exposure time. mA is inversely proportional to the exposure time. *e.g.*

  300 mA at 1/60 sec. = 5 mAs

  200 mA at 1/40 sec. = 5 mAs.

2. kVp

- kVp is the voltage applied between the cathode and anode.
- Increase in kVp increases the positive charge on the anode and thus accelerate the electrons towards the anode (target). Thus an X-ray beam of short wavelength and more penetrating power is produced.
- High kVp required for thicker part.

- Higher kVp gives a longer scale of contrast and more exposure latitude (degree of variation from the correct exposure factors). And hence used for soft tissue examination.
- Increase in kVp increases radiographic density because more number of x-ray penetrates the patient and exposes the film.
- Adequate radiographic density is maintained by-

Adding kV for each cm increase in body thickness	When the original kV is below
2	80s
3	80-100
4	>100

- When kVp is increased by 20% radiographic density is doubled.
- When kVp decreased by 16 % radiographic density is halved.

Exposure latitude: Degree of variation from the correct exposure factors, which still produces a diagnostic radiograph. It varies with the range of kVp used.

- High kVp range technique increases the exposure latitude.

kVp range	Exposure latitude
46-55	±2 kVp
56-65	±4 kVp
66-75	±6 kVp
76-85	±8 kVp
86-95	±10kVp

- High kVp range produces a long scale contrast which is desirable for most radiographic examinations.

A change in kVp requires a change in mAs and vice-versa to maintain similar radiographic density.

kVp range	kVp change required when mAs doubled or halved
41-50	±4
51-60	±6
61-70	± 8
71-80	±10
81-90	±12
91-100	±14
101-110	±6

3. Focal film distance (FFD)

- Distance between the target and film is known as FFD.
- Appropriate FFD for most procedures is 36-40 inches.

- Inverse square law: The intensity of x-ray beam is inversely proportional to the square of the distance from the target.

$$\frac{Old\,mAs}{New\,mAs} = \frac{(New\;FFD)^2}{(Old\;FFD)^2}$$

- If FFD is doubled, the mAs must be increased 4 times to maintain radiographic density.
- FFD change does not affect the penetrating power of the x-ray beam because kVp remains constant.

4. Object film distance (OFD)

- Distance between the object being imaged and radiographic film.
- OFD should be short to minimize the penumbra and less magnification.

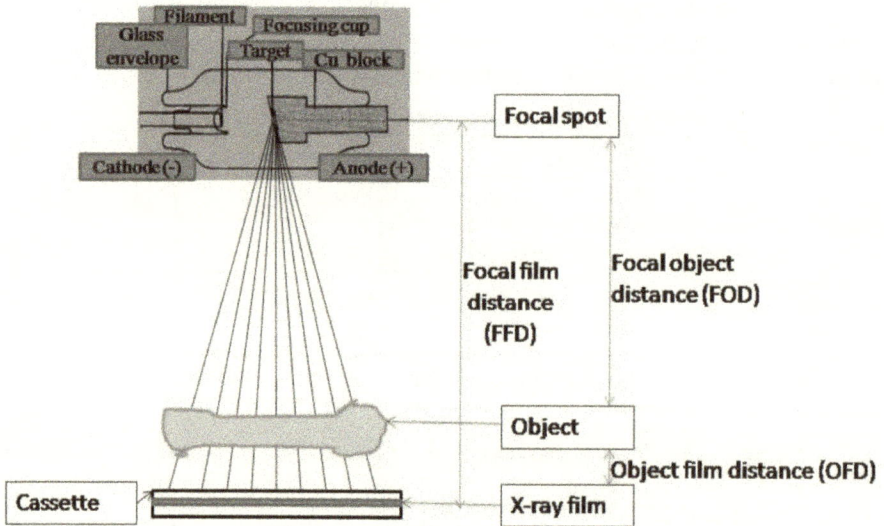

Fig. 6.1: Measurement of FFD, FOD and OFD

7

Radiographic Positioning

Correct positioning is essential part of the radiography to obtain a radiograph of accurate details of the part being examined.

- It refers to placement of body part in respect to linear alignment of X-ray tube and film.
- Misinterpretation can result from inaccurate positioning.

Patient preparation and restraining:

- Remove the dust, mud, feces, metal or leather straps, metal chain etc. before taking a radiograph.
- Fast the animal; give enema and gas absorbing preparations (charcoal) before radiography of abdominal cavity.
- Restrain the animal either by physical restrains or by sedatives or perform anesthesia if animal is aggressive.

Terminology used for direction

Rostral (R): area on the head situated towards the nose.

Palmer (Pa): Situated on the caudal aspect of the front limb, distal to the antebrachio-carpal joint.

Planter (Pl): Situated on the caudal aspect of the rear limb, distal to the tarsocrural joint.

Proximal (Pr): Situated closer to the point of attachment or origin.

Distal (Di): Situated away from the point of attachment or origin.

- Radiographic positioning is based on point of entrance to point of exit of the X-ray beam in the body part.
- Cassette having X-ray film is placed below/posterior/caudal/palmer/planter to the point of exit of X-ray beam.

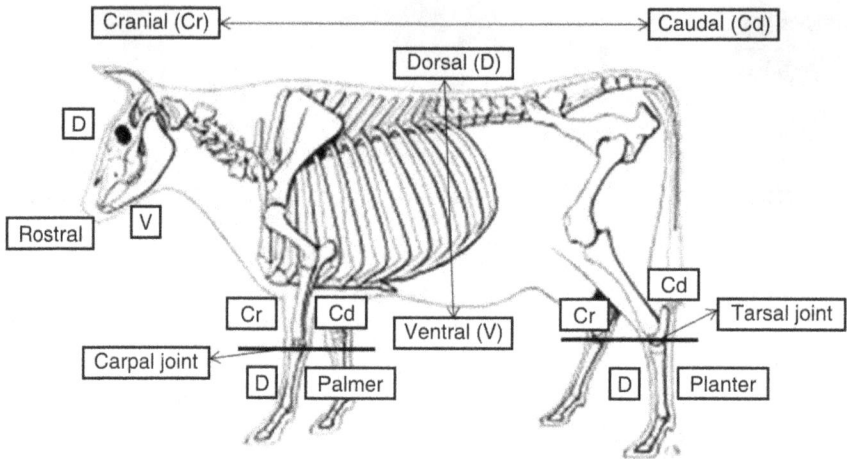

Fig. 7.1: Terminology used for direction

Fore limbs

1. **Scapula:** Caudocranial, Lateral (mediolateral)
2. **Shoulder joint:** Caudocranial,Lateral (mediolateral)
3. **Humerus:** Mediolateral, Caudocranial
4. **Elbow joint:** Mediolateral, Lateromedial (used for fragmented coronoid process), Craniocaudal
5. **Radius-ulna:** Mediolateral, Craniocaudal
6. **Carpus, Metacarpus and Digit:** Mediolateral, Dorsopalmer.

Hind limbs

1. **Femur:** Mediolateral and Craniocaudal
2. **Stifle joint:** Mediolateral, Caudocranial and sunrise view
3. **Tibia:** Mediolateral, Caudocranial
4. **Tarsus, Metatarsus and Digit:** Mediolateral, Plantarodorsal

Skull: Lateral and Ventrodorsal

1. Maxilla: Ventrodorsal open mouth
2. Tympanic bulla: Open mouth
3. Mandible: Ventrodorsal and Lateral
4. Maxillary and mandibular dental arcade: Lateral oblique

Vertebral column

1. Cervical vertebrae: Lateral, Ventrodorsal, Dorsoventral, Oblique, Flexed lateral, Extended lateral
2. Thoracic, Lumbar and sacral: Lateral and Ventrodorsal

Pelvis: Lateral, ventrodorsal flexed position, Ventrodorsal extended position

Abdomen: Lateral and Ventrodorsal (*Take radiograph at peak expiration*).

Thorax: Lateral and Ventrodorsal and Dorsoventral (*Take radiograph at peak Inspiration*).

S.No.	View	Position of animal	Point of entrance of X-ray	Point of exit of X-ray
1.	Dorso-Ventral (DV)	Sternal/ventral	Dorsal (D) recumbency	Ventral (V)
2.	Ventro-Dorsal (VD)	Dorsal recumbency or supine position	Ventral (V)	Dorsal (D)
3.	Cranio-Caudal view of limbs	Limb extended forward	Cranial (Cr)	Caudal (Cd)
4.	Right Dorsal-Left Ventral oblique view (RD-LV)	Ventral recumbency	Right Dorsal	Left Ventral
5.	Lateral	Lateral	Lateral (L) / Medial (M)	Medial (M) / Lateral (L)
6.	Flexor view of bone in horses navicular	Standing	Beam parallel to flexor surface	Towards the film

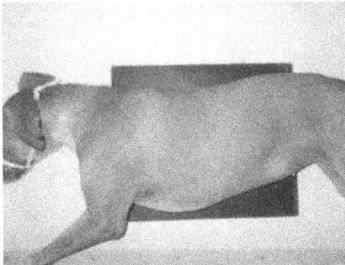

Fig.7.2a: Positioning for lateral image of the thorax

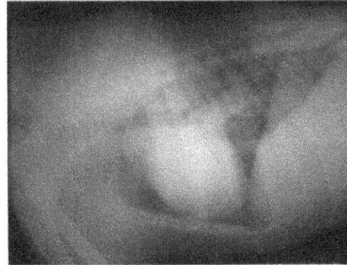

Fig.7.2b: Lateral image of the thorax

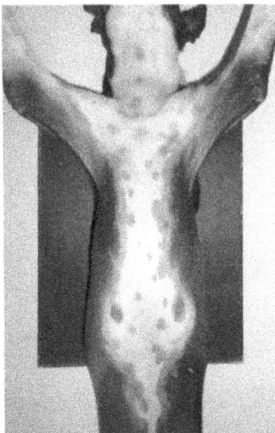

Fig. 7.3a: Positioning for ventro-dorsal image of the thorax

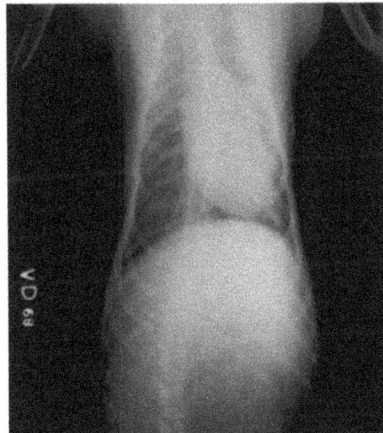

Fig. 7.3b: Ventro-dorsal image of the thorax

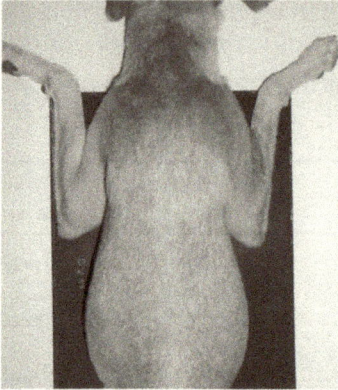

Fig. 7.4a: Positioning for dorso-ventral image of the thorax

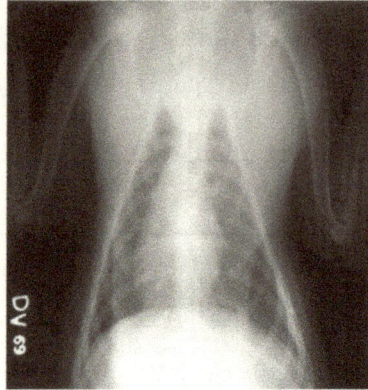

Fig. 7.4b: Dorso-ventral image of the thorax

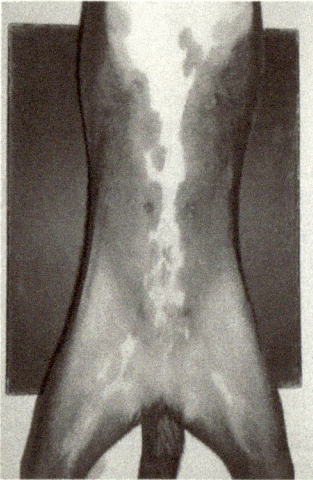

Fig. 7.5a: Positioning for ventro-dorsal image of the abdomen

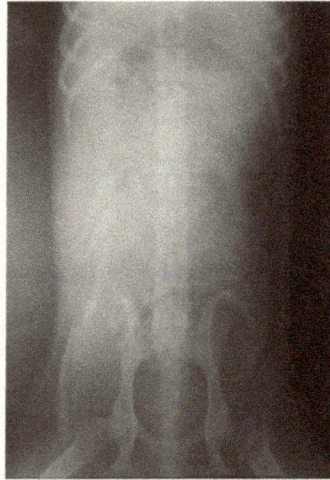

Fig. 7.5b: Ventro-dorsal image of the abdomen

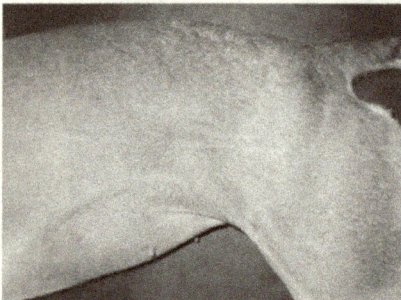

Fig.7.6a: Positioning for lateral image of posterior abdomen and pelvic cavity

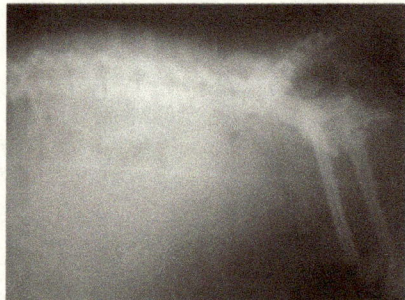

Fig. 7.6b: Lateral image of posterior abdomen and pelvic cavity

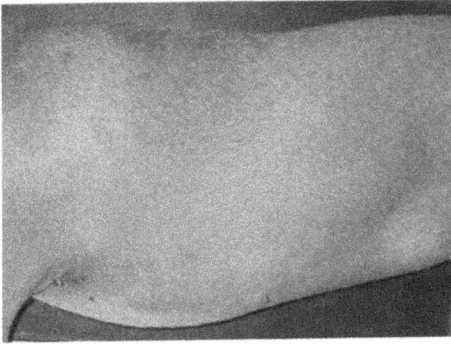

Fig. 7.7a: Positioning for lateral image of abdomen

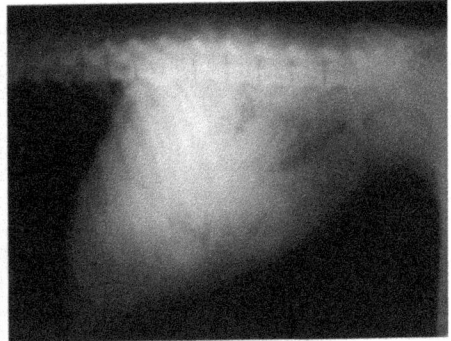

Fig. 7.7b: Lateral image of abdomen

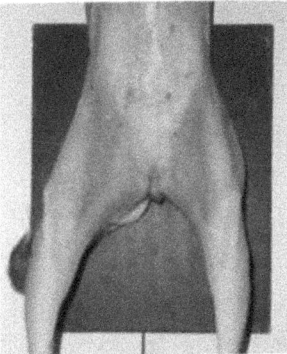

Fig. 7.8a: Positioning for ventro-dorsal projection of the pelvis

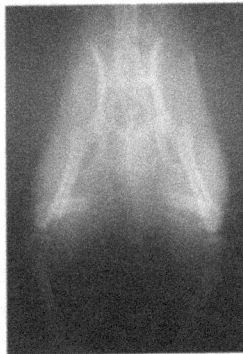

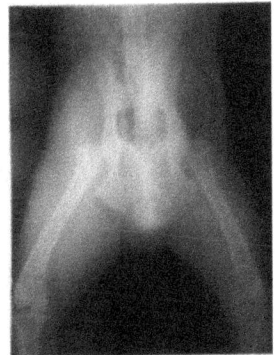

Fig. 7.8b: Ventro-dorsal projection of the pelvis

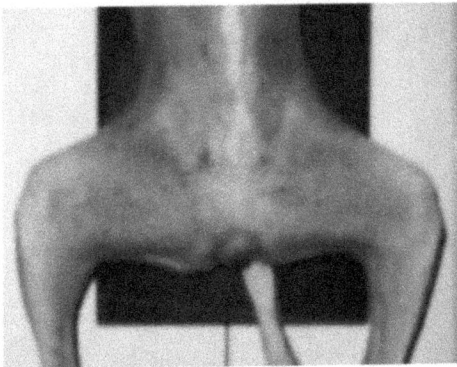

Fig. 7.9a: Positioning for ventro-dorsal projection of the pelvis in flexed position

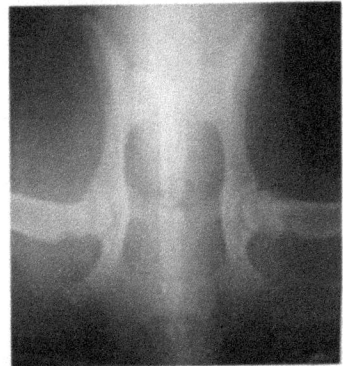

Fig. 7.9b: Ventro-dorsal projection of the pelvis in flexed position

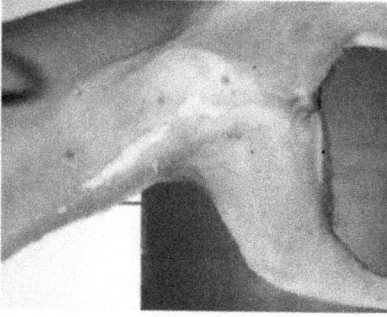

Fig. 7.10a: Positioning for medio-lateral image of the femur

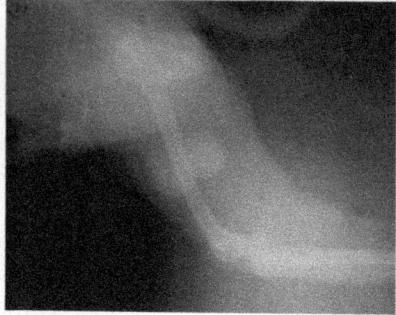

Fig. 7.10b: Medio-lateral image of the femur

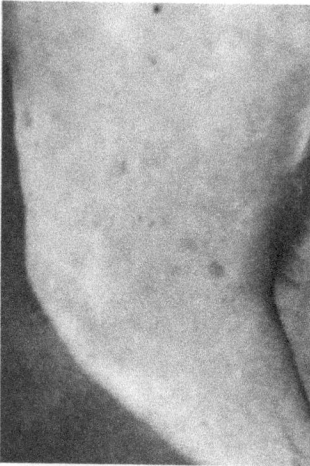

Fig. 7.11a: Positioning for medio-lateral image of the stifle joint

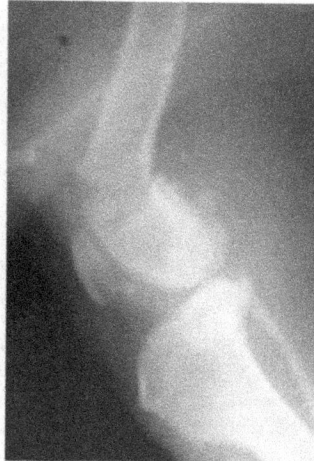

Fig. 7.11b: Medio-lateral image of the stifle joint

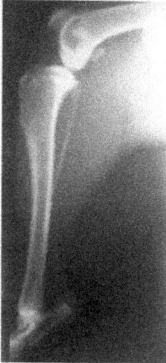

Fig. 7.12a: Positioning for medio-lateral image of the Tibia

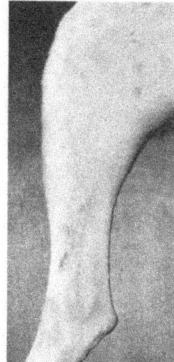

Fig. 7.12b: Medio-lateral image of the Tibia

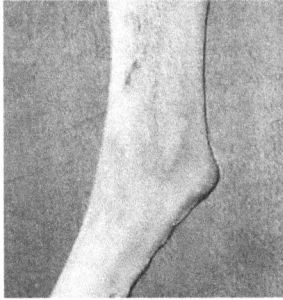

Fig. 7.13a: Positioning for medio-lateral image of the Hock joint

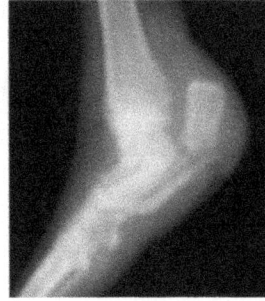

Fig. 7.13b: Medio-lateral image of the Hock joint

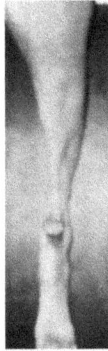

Fig. 7.14a: Positioning for planto-dorsal view of the tarsal joint

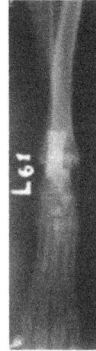

Fig. 7.14b: Planto-dorsal view of the tarsal joint

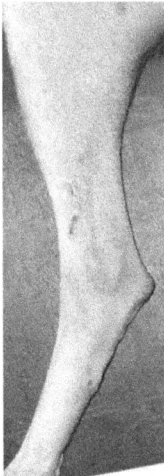

Fig. 7.15a: Positioning for Medio-lateral view of the tarsal joint

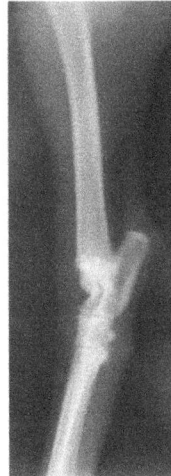

Fig. 7.15b: Medio-lateral view of the tarsal joint

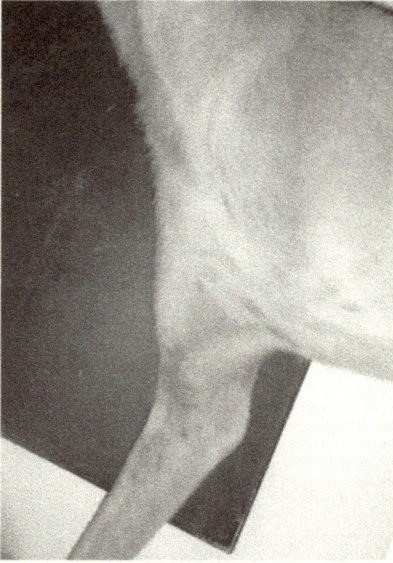

Fig. 7.16a: Positioning for medio-lateral projection of humerus

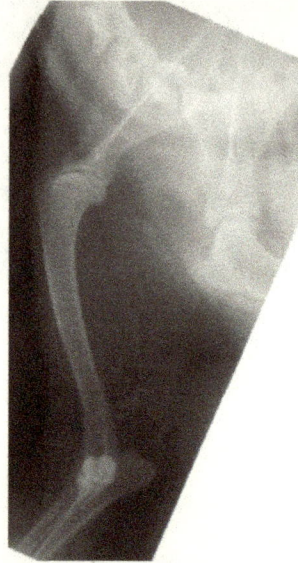

Fig. 7.16b: Medio-lateral projection of humerus

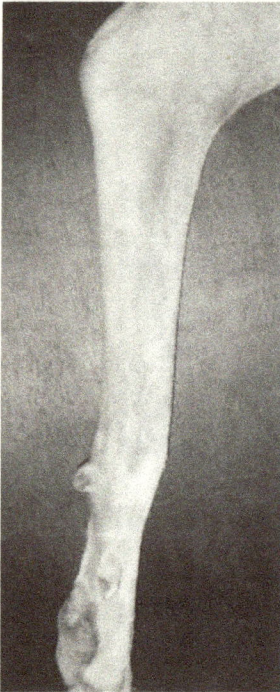

Fig. 7.17a: Positioning for medio-lateral projection of radius-ulna

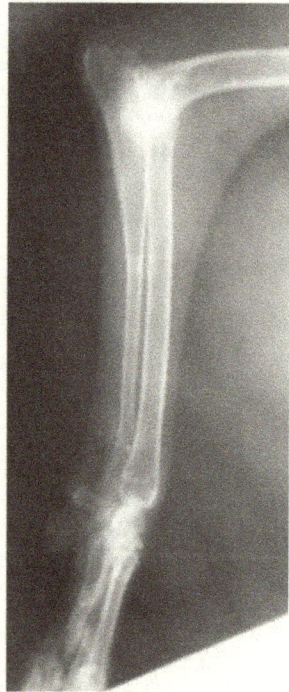

Fig. 7.17b: Medio-lateral projection of radius-ulna

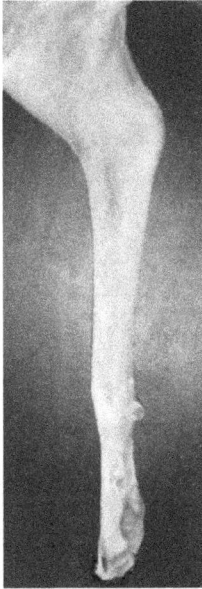

Fig. 7.18a: Positioning for medio-lateral projection of carpus, metacarpus and phalanges-ulna

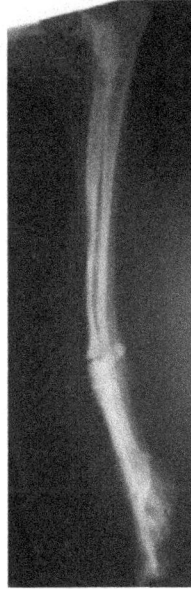

Fig. 7.18b: Medio-lateral image of carpus, metacarpus and phalanges

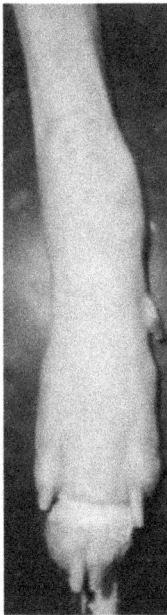

Fig. 7.19a: Positioning for dorso-palmer projection of carpus, metacarpus and phalanges-ulna

Fig. 7.19b: Dorso-palmer image of carpus, metacarpus and phalanges

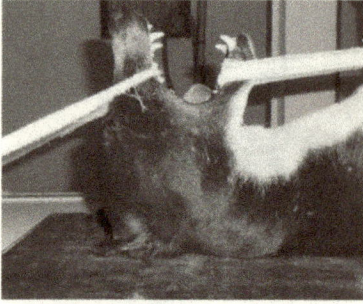

Fig. 7.20a: : Positioning of head for an open mouth tympanic bulla projection

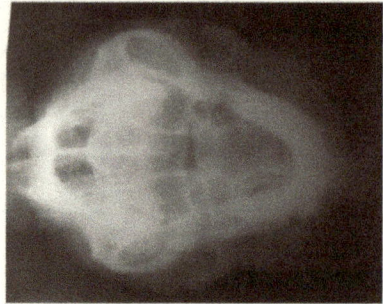

Fig. 7.20b: Open mouth tympanic bulla projection

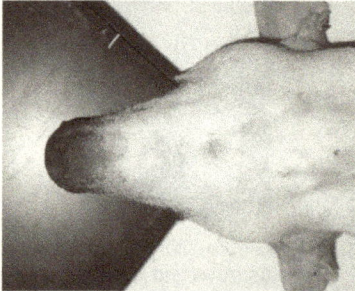

Fig. 7.21a: : Positioning for intraoral image of the mandible

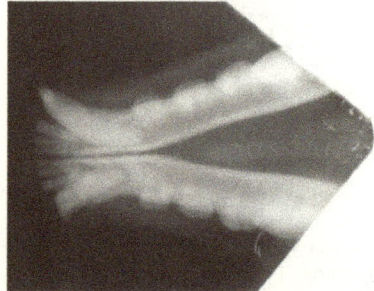

Fig. 7.21b: Intraoral radiographic image of the mandible

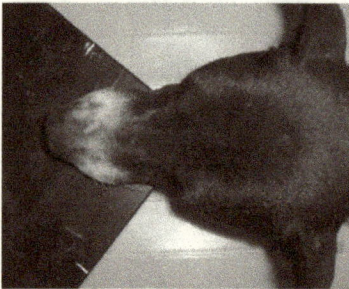

Fig. 7.22a: : Positioning for intraoral image of the maxilla

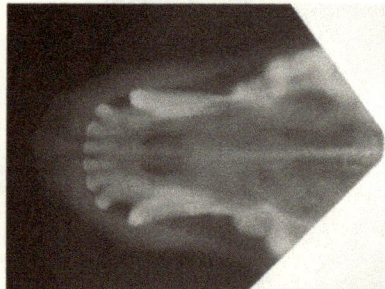

Fig. 7.22b: IIntraoral radiographic image of the maxilla

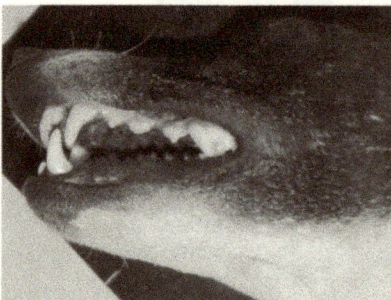

Fig. 7.23a: : Positioning for a lateral oblique view of maxillary dental arcade

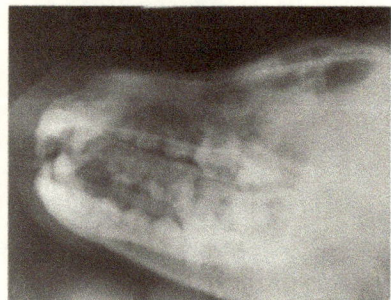

Fig. 7.23b: Lateral oblique view of maxillary dental arcade

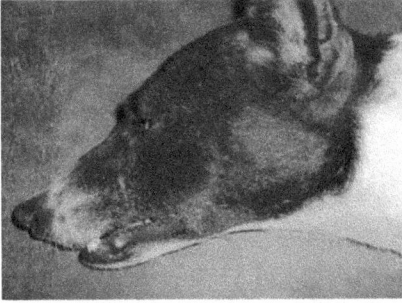

Fig. 7.24a: Positioning for a lateral oblique view of maxillary dental arcade

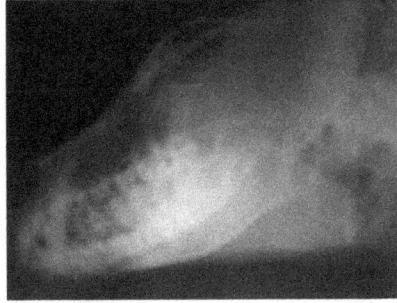

Fig. 7.24b: Positioning for a lateral oblique view of maxillary dental arcade

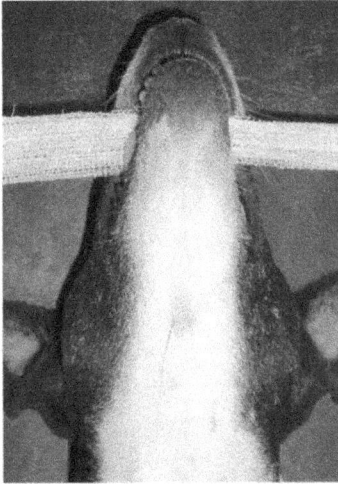

Fig. 7.25a: Positioning for ventro-dorsal view of the skull

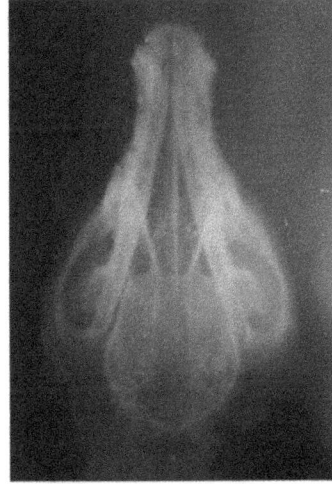

Fig. 7.25b: Ventro-dorsal view of the skull

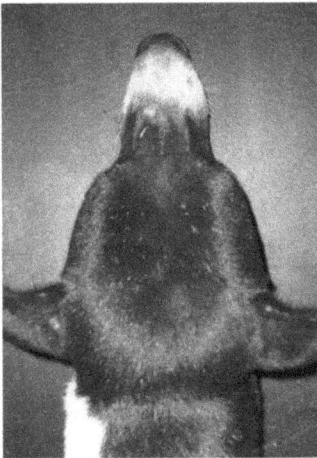

Fig. 7.26a: : Positioning for dorso-ventral view of the skull

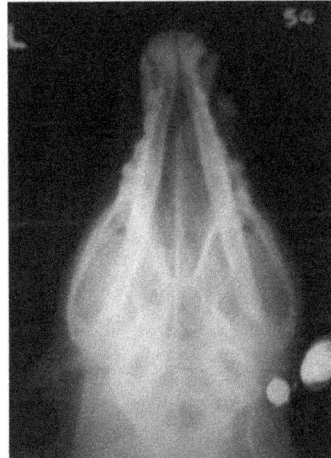

Fig. 7.26b: Dorso-ventral view of the skull

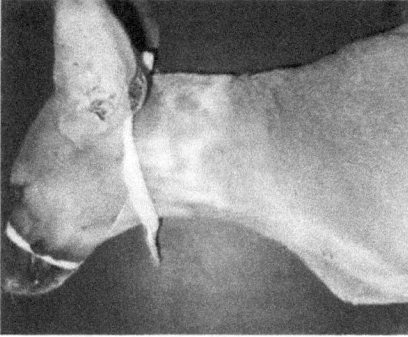

Fig. 7.27a: : Positioning for lateral view of the cervical vertebrae

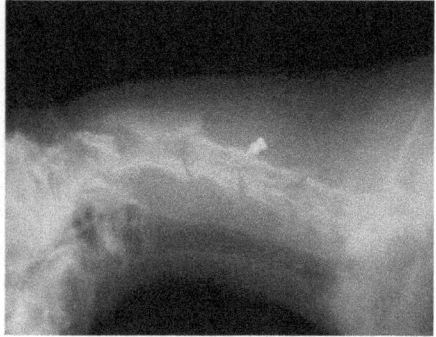

Fig. 7.27b: Lateral view of the cervical vertebrae

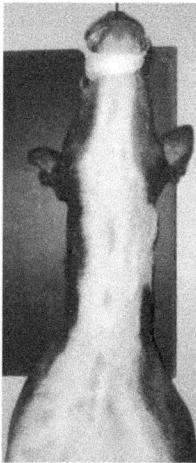

Fig. 7.28a: : Positioning for ventro-dorsal view of the cervical vertebrae

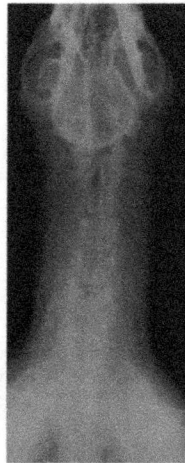

Fig. 7.28b: Ventro-dorsal view of the cervical vertebrae

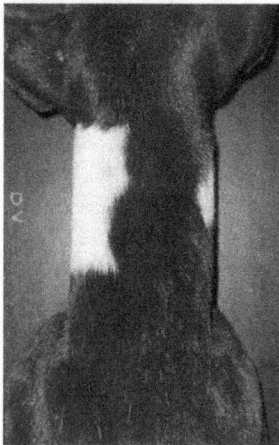

Fig. 7.29a: : Positioning for dorso-ventral view of the cervical vertebrae

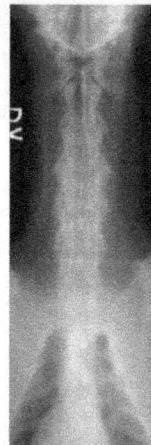

Fig. 7.29b: Dorso-ventral view of the cervical vertebrae

8

Developing a Small Animal Radiographic Technique Chart

Development of radiographic technique chart is essential for the production of diagnostic radiograph of good quality.

- Exposure technique for a thorax measuring-
    - (i) 20 Cm: 4.8mAs and 80 kVp
    - (ii) 26 Cm: 9.6mAs and 80 kVp
    - (iii) 30 Cm: 19.2mAs and 80 kVp
- kVp and mAs radiographic density relationship:
    1. **kVp**
        - (i) Increase the kVp by 20% - Double the radiographic density.
        - (ii) Decrease the kVp by 16% - Halve the radiographic density.
    2. **mAs**
        - (i) Double the mAs-Double the radiographic density.
        - (ii) Halve the mAs-Halve the radiographic density.
- Use of grid: With the high speed screens

S.No.	Technique	mAs
1.	Table top abdomen	3.3
2.	Grid abdomen	9.9
3.	Table top thorax	1.6
4.	Grid thorax	4.8
5.	Table top bone	4.3
6.	Grid bone	12.9

Santes' rule: A method of estimating kilovoltage in relation to area thickness: (2 x thickness in cm) + 40 = kVp.

9 X-ray Films, Fluorescent Screens and Intensifying Screens

X-RAY FILMS

X-ray films are of two types

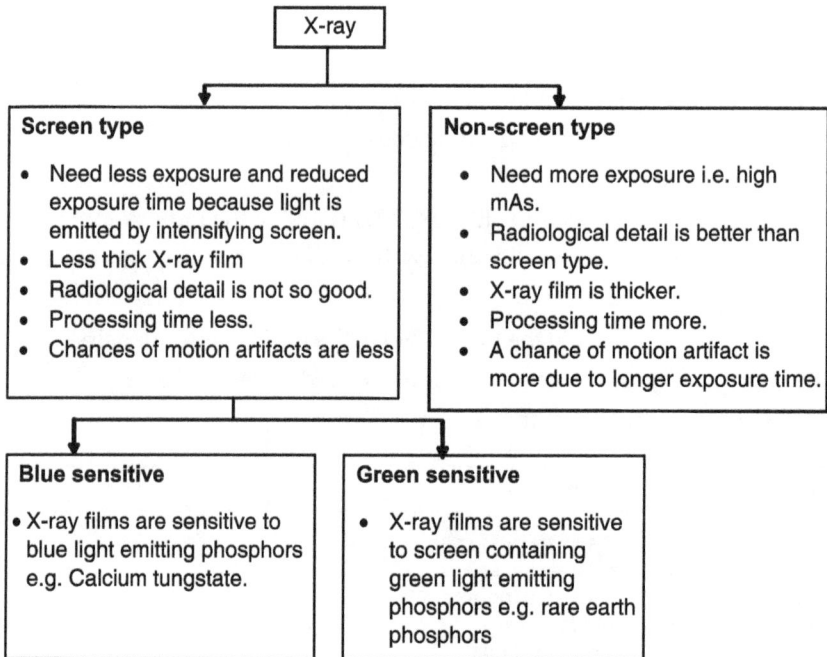

```
                    ┌─────────┐
                    │  X-ray  │
                    └─────────┘
           ┌──────────────┴──────────────┐
```

Screen type

- Need less exposure and reduced exposure time because light is emitted by intensifying screen.
- Less thick X-ray film
- Radiological detail is not so good.
- Processing time less.
- Chances of motion artifacts are less

Non-screen type

- Need more exposure i.e. high mAs.
- Radiological detail is better than screen type.
- X-ray film is thicker.
- Processing time more.
- A chance of motion artifact is more due to longer exposure time.

Blue sensitive

- X-ray films are sensitive to blue light emitting phosphors e.g. Calcium tungstate.

Green sensitive

- X-ray films are sensitive to screen containing green light emitting phosphors e.g. rare earth phosphors

X-Ray film sizes: 5 x 7, 6.5 x 8.5, 6 x 12, 6 x 15, 8 x 10, 10 x 12, 11 x 14, 12 x 12, 12 x 15, 14 x 14, and 14 x 17.

Other types of the films

1. *Automatic processor films:* these films are hard and used in automatic processor.
2. *Occlusal films, Bitewing Film and Periapical Film:* Used for intraoral radiography.

Composition of screen type X-ray film: It is double coated *i.e.* emulsion coating on both side of the base. X-ray film is consisting of three layers each on both side of the film base. Total thickness of the film is 0.25 mm.

1. **Film base**
   - The center of the film which provides support is called base
   - It is impermeable, rigid and water resistant.
   - The base is made up of Polyester (0.007" thick) or Cellulose triacetate (0.008" thick).
   - It gives blue tint to the film.
2. **Adhesive:** It is used to achieve firm contact between emulsion and base.
3. **Emulsion layer**
   - It contains finely precipitated silver halide crystals in a gelatin base. Silver halide means:
     - (*i*) Silver bromide crystals (90-99%).
     - (*ii*) Silver iodide crystals (1-10%).
   - The diameter of halide crystals: 1-1.5μ.
   - Number of halide crystals in the emulsion: 6.3×10^9 crystals/cu mm of emulsion.
   - Thickness of emulsion layer: 0.0005"
   - Film speed is decided by crystal size. Larger the crystal size, higher the speed.
4. **Protective layer:** It is the thin layer of gelatin spread over the emulsion layer. It protects emulsion from mechanical damage.

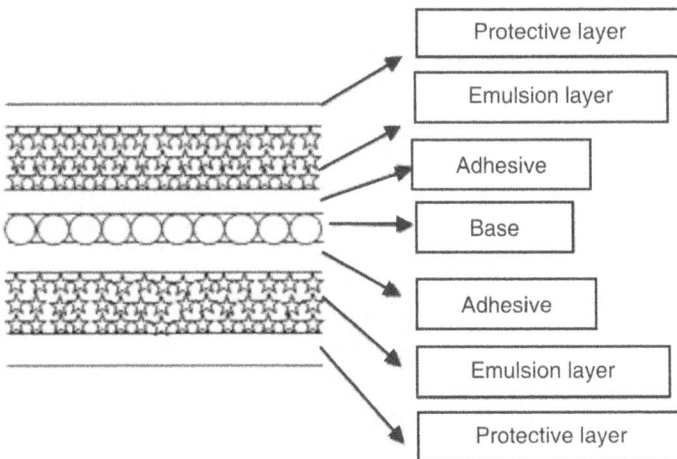

Fig. 9.1: Layers of X-ray film

Film speed: On the basis of film speed film can be divided in to-
1. High or regular or fast
2. Average or par
3. Slow

- Fast speed films are more sensitive and require lower mAs.
- High speed films require less exposure than slow speed film to produce a given radiographic density.
- High speed film produces image of granular appearance (decreased detail).
- Par speed films are used in veterinary practice.

Film latitude

- It is the film's inherent ability to produce shades of gray.
- Film latitude is increased; image of long scale of contrast is produced.

FLUORESCENT SCREENS

These screens contains

1. Calcium tungstate
   - Used in intensifying screens.
   - Emits blue fluorescence.
2. Zinc sulphide
   - Used in high speed intensifying screens.
   - Emits violet-blue fluorescence.
3. Barium platinocyanide
   - Emits violet-blue fluorescence.
4. Silver activated zinc cadmium sulphide
   - Used in fluoroscopic screen and image intensifier.
   - Emits yellow green fluorescence. Eye is more sensitive to this colour.
5. Rare earth screens
   - Examples are Terbium activated gadolinium oxysulfide and Lanthanum oxybromide. Both emits green light.
   - Used for very fast screens.

INTENSIFYING SCREENS

- These screens contain fluorescent crystals. The crystals emit foci of light when exposed to X-rays.
- 95% of film density is caused by fluorescence of intensifying screen whereas 5% by the direct X-ray exposure.
- The film is sandwiched between two intensifying screens mounted in a cassette.
- Thickness of front screen is 1.25mm and back screen is 1.5mm.
- An intensifying screen has four layers.

1. **Base:** Made of plastic or cardboard.
2. **Reflecting layer**
   - This layer reflects the light back towards the film side or front of screen.
   - Made of coating of $BaSO_4$, MgO and titanium dioxide.
3. **Phosphor layer**
   - It converts the X-ray energy into visible light
   - Made of calcium tungstate.
4. **Protective layer**
   - Consist of cellulose compound.

Film screen combinations: Classification of film screen system-

100 speed system	200 speed system	400 speed system
• Produces excellent detail • Require highest exposure	• Produces good detail • Two times faster than 100 speed system	• Produces good balance of speed and detail. • Mostly used in Veterinary practice

> **A slow speed film produces an image with excellent detail but requires higher exposure**

Care of intensifying screen

- Regular cleaning of screens with intensifying screen cleaning solution or warm water.
- Don't use denatured alcohol or abrasive products.

Screen speed: Screen speed depends on size of phosphor crystals. High speed screen have larger crystals.

1. **High (regular):** Requires less exposure as compared to par and slow speed screens but detail is decreased.
2. **Par (medium):** mAs need to be increased 2 times from high speed to maintain radiographic density.
3. **Slow (fine):** require longer exposure time and detail is better. mAs need to be increased 4 times.

RADIOGRAPHIC DENSITY

Degree of blackness on the radiograph is known as radiographic density. X-rays passes through the patient and exposed the emulsion of the film. After processing these areas appear dark due to deposits of black metallic silver.

Factors affecting radiographic density:

- **mAs:** Increased by increasing either mA or exposure time (by increasing number of electrons travelling from cathode to anode).
- **kVp:** Increased kVp increases radiographic density (by increasing penetration power of the beam).
  - At low kVp range: Small variation in kVp causes more change in radiographic density.
  - At high kVp range: Small variation in kVp causes less effect on radiographic density.
- **Subject density:** Subject density is the weight/unit volume of different body tissues and inversely proportional to the radiographic density.
  - *(i)* The tissue which allows X-rays to pass through them appears blacker on the processed film - radiolucent (e.g. normal lungs).
  - *(ii)* The tissues through which most of X-rays can not pass appear whiter - radio opaque (e.g. bones).
- **FFD:**

$$\text{Radiographic density} \propto \frac{1}{(FED)^2}$$

- **Dark room processing:** increased developing time or temperature or both leads to increased radiographic density.

RADIOGRAPHIC CONTRAST

Difference in radiographic density between adjacent areas on a radiograph is known as radiographic contrast.

1. **Long or low scale of contrast:** (at high kVp) few black and white shades with many shades of gray i.e. image with many densities.

2. **Short or high scale of contrast:** Black and white shades predominates with few shades of gray in between.

Factors affecting radiographic contrast

- **Subject density:** Ability of different body tissues to absorb X-rays.
    - Air (Least dense) < Fat < Water or muscle < Bone<Metal (Most dense).
    - Penetration power of X-rays depends on atomic number and thickness of tissues.
    - Bone absorbs more X-rays than muscle
- **kVp:** Radiographic contrast is directly proportional to kVp. At high kVp more exposure latitude.
- **Film contrast:** It also affects the radiographic contrast.
- **Film fogging:** It decreases the radiographic contrast.
- **Secondary radiation:** Scatter radiation affects the radiographic contrast.
- **Dark room processing:** A decrease in radiographic contrast is noted with decreasing development time, use of old developing solution and more warm solution.

RADIOGRAPHIC DETAIL

Radiographic detail is characterized by sharp tissue and organ interfaces. For a good diagnostic radiograph, the radiographic detail should be good.

Factors affecting radiographic detail:

- **Motion of the patient and long exposure time:** Motion of the patient leads to loss of radiographic detail. To overcome this problem sedate the patient and use shorter possible exposure time.
- **Penumbra effect:** The following factors affects the penumbra effect
    - **Focal spot size:** Increase in focal spot size increases penumbra.
    - **Focal film distance (FFD):** Increase in FFD decreases penumbra. But FFD can not be increased above a limit because of inverse square law (intensity decreases at a rate inversely to the square of the distance) e.g. If FFD is doubled; the mA should be increased 4 times. For minimizing penumbra a FFD of 36-40 inches should be maintained.
- **Object film distance (OFD):** It is the distance between the film and the object to be imaged. Decrease in OFD decreases penumbra. So the affected side should be placed near to film.
- **Film intensifying screen contact:** If the distance between the film and intensifying screen within cassette is more, fluorescence will cause more reflection forming a blurred image

- Better detail is obtained by using non screen film as compared to screen films.

- Overexposure and underexposure affects the radiographic detail.

- Dark room processing is equally important. Different time and temperature combinations can be used for better detail e.g. at 68°F for 5 minutes.

RADIOGRAPHIC MOTTLE

Density variation in a radiograph made with intensifying, given a uniform exposure. It is the result of quantum mottle, structure mottle and film graininess.

1. *Quantum mottle*
   - X-rays are produced in small packets of energy called quanta/ photons.
   - Quantum mottle is a density variation on finished radiograph due to random spatial distribution of X-ray quanta absorbed in the intensifying screen.
   - If faster speed film and intensifying screen are used, fewer x-rays are required to produce a given radiographic density and therefore quantum mottle is more noticeable because of the fewer number of X-rays needed to produce a given density.

2. *Structure mottle*
   - Fluctuation in the density due to non uniform structure of intensifying screen.
   - Film graininess: Random distribution of developed silver deposits leads to film graininess.

IMAGE DISTORTION

- **Foreshortening:** occurs when the body part is not parallel to the recording surface (Fig. 10.1a).
- As the OFD is increased, penumbra is increased and size of the projected image is also magnified (Fig. 10.1b).
- If X-ray beam is not perpendicular to the recording surface, distortion occurs.

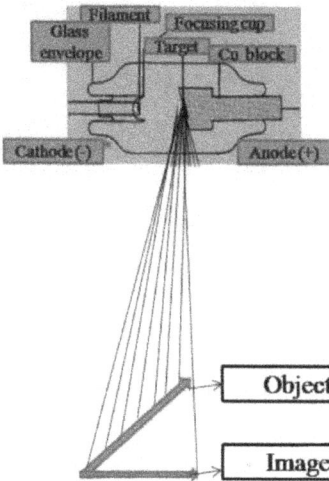

Fig. 10.1a: Foreshortening

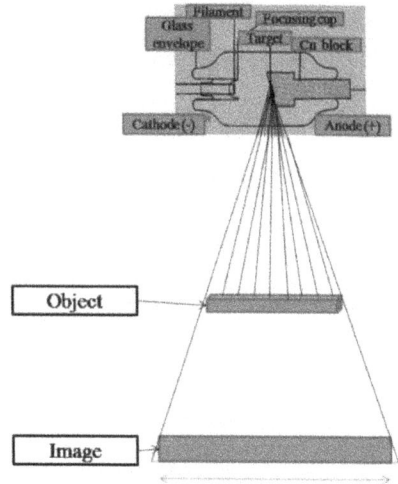

Fig. 10.1b: Image magnification

Image magnification

- Shorter the distance between the object and light source, greater will be the magnification of image (Fig. 10.1b).

$$\% \text{ Magnification} = \frac{\text{Width of image} - \text{Width of object}}{\text{Width of object}} \times 100$$

$$\text{Magnification factor} = \frac{\text{FFD}}{\text{FOD}}$$

- To reduce magnification, part to be examined should be as close to the film as possible.
- FFD should be kept 90-100 cm.

11

When an X-rays photon strikes on an object it

(*i*) Penetrates the object

(*ii*) Absorbed

(*iii*) Scattered.

- It produces fog on the film (decreased contrast).
- Because of long wave length (less energy) than primary beam scatter radiation projected in all direction and hazardous to the patient and personnel.
- Scatter radiation is produced when kVp is increased.
- Part thickness > 9cm produces scatter radiation of such quality that can decrease the detail of the radiograph.

X-RAY BEAM RESTRICTORS

Restricts the primary X-ray beam and reduces scatter radiation. Common X-ray beam restrictors are:

1. **Cone:** are made up of lead (Pb) and placed over the collimator at the tube window. It restricts the primary beam and is marketed in fixed sizes so it must be changed for different area to be radiographed.

2. **Diaphragm:** It is the lead (Pb) sheet with rectangular, circular or square hole (fixed size) in the center. It also restricts the primary beam.

3. **Collimators:** Consist of adjustable lead shutters in the tube head that can be adjusted on the basis of area being examined. The advantage is that it has a light that illuminates the area to be examined, *i.e.* actual size of the primary beam field.

4. **Filters:** A thin sheet of aluminum is known as filter which is placed over the tube window to absorb the soft (less energetic and less penetrating) X-rays. Filter reduces the radiation hazard to the patient and personnel. The advantage is that it reduces scatter radiation and thus increases radiographic detail.

GRIDS

A grid is a series of thin, linear strips of alternating radiodense material (lead) and radiolucent interspaces (plastic, aluminum or fiber).

- Use: To decrease the scatter radiation and increase the film contrast.
- Why use of grid: High kVp technique is used to make radiograph of thick areas and thus more scatter radiation is produced causing loss of detail on radiograph. To eliminate the scatter radiation, grids are used with areas measuring > 9cm.
- The grid is placed between the patient and the cassette to limit the primary beam.
- When X-ray penetrates the patient and most of the primary beam passes through the lead strips to expose the film. However, the scatter radiation is absorbed by lead strips (Fig. 11.1a).

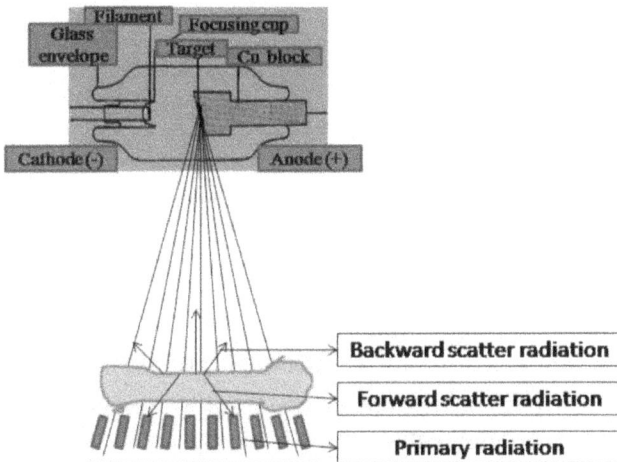

Fig. 11.1a: X-ray penetrates the patient and most of the primary beam passes through the lead strips to expose the film. However, the forward scatter radiation is absorbed by lead strips.

- Some usable X-rays are also absorbed by grid, to compensate for this loss, the no. of X-rays generated must be increased by increasing mAs (up to 6.6 times from the normal mAs) and kVp.

Grid ratio

- The grid ratio is the ratio of the height of the lead strips to the width between 2 lead strips.

  e.g. Height of the lead strip = 2.5mm

  Width between lead strip = 0.5 mm

  $$\text{Grid ratio will be} = \frac{\text{Height of the lead strip}}{\text{Width between lead strip}} = \frac{2.5}{0.5} = 5{:}1$$

- Grid efficiency is controlled by grid ratio. If grid ratio is higher, grid efficiency is more i.e. 10:1 grid is more efficient than 5:1
- High ratio grid absorbs more scatter radiation and primary beam. So mAs should be increased as the grid ratio increases to maintain the radiographic density.

Grid ratio	Increase in mAs requires
5:1	3 times
8:1	4 times
12:1	5 times
15:1	6 times

Types of Grid

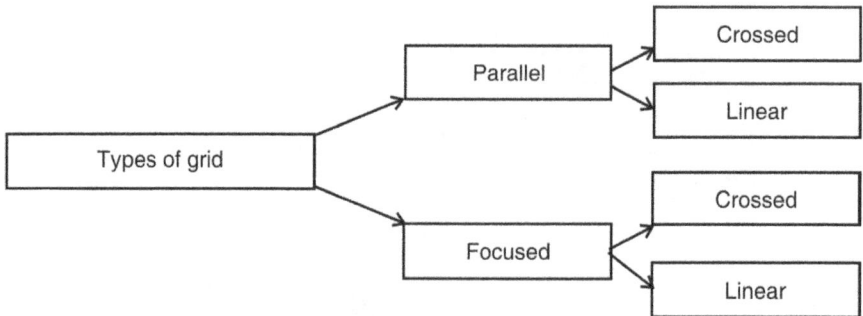

1. Parallel grid

- Lead stripes are placed parallel to the grid surface (**Fig. 11.2b**).
- Lead strips primarily absorb scatter radiation whereas primary beam passes through the interspaces to expose the film.
- Disadvantages:
  - The primary X-ray beam is absorbed at the periphery of the grid and thus decreasing amount of X-rays reaches the film near the grid edges this is known as *grid cut off*.
  - Grid cut off is more with grids having higher grid ratio.

Fig. 11.2b: Parallel grid

2. Focused grid

- To overcomex the grid cut off problem, focused grid were developed.
- The lead stripes are placed with progressively increasing angles to match the angle of primary X-ray beam (**Fig. 11.2c**).

- *Disadvantages:* The primary beam is nearly completely absorbed except at the centre.
  - If the grid is not perpendicular
  - If the grid is not properly centered.
  - Upside down grid.

Fig. 11.2c: Focused grid

3. Linear grids

- Lead strips are parallel to each other (**Fig. 11.2d**).
- The grid is placed in such a way that the lead stripes should be parallel to the length of the table.
- The primary X-ray beam can be angled along the length of grid without grid cut off.

Fig. 11.2d: Linear grid

4. Crossed grid

- Two linear grids placed at right angle (**Fig. 11.2e**).
- Removes more scatter radiation than the linear grid.
- The tube cannot be tilted without producing grid cut off.
- High mAs is required as compared to linear grid due to more no. of X-rays are absorbed by crossed grid (because of more lead).

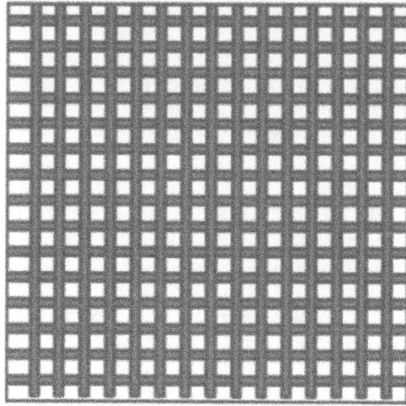

Fig. 11.2e: Crossed grid

Grid cut off: Grid cut off is present if the grid is

(a)   Tilted or off level (**Fig. 11.2f**)

(b)   Upside down (**Fig. 11.2g**)

(c)   Off centered (**Fig. 11.2h**)

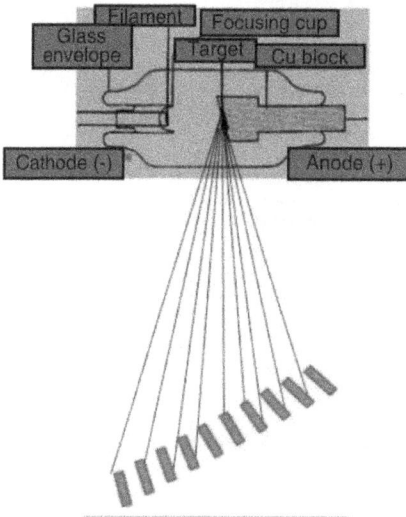

Fig. 11.2f

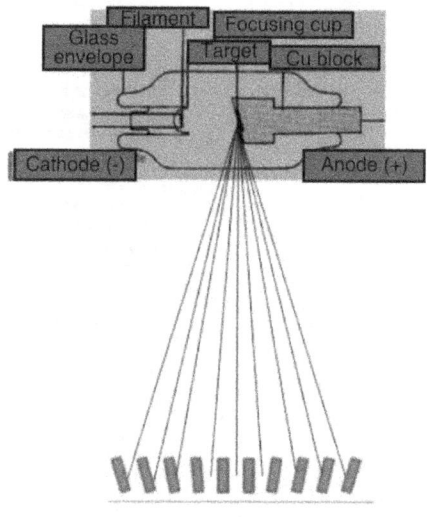

Fig. 11.2g

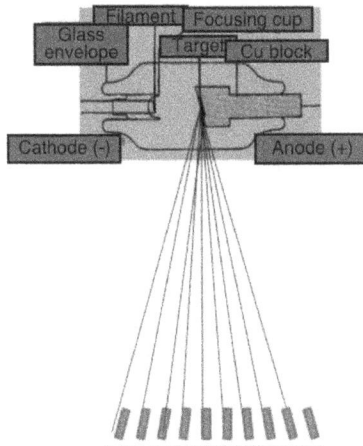

Fig. 11.2h

Fig. 11.2: Grid cut off: **f** Tilted or off level grid **g.** Upside down grid **h.** Off centered grid

Grid lines: Grid produces very thin white lines on the radiograph.

- Grid lines can be decreased by-
    - (*i*) Thinnest lead strips with effective absorption of scatter radiation.
    - (*ii*) By increasing the number of lead lines/inch (increased grid ratio and hence requires higher mAs).
    - (*iii*) Use of potter bucky diaphragm or Bucky. The bucky is placed under the X-ray table with a tray to hold the cassette.
- Bucky sets the grid in motion.
- Disadvantage: Noise and vibration are produced during motion of the bucky which may disturb the animal during exposure.
- Reciprocating devices have advantages *e.g.* it provides constant motion of the grid, shortened exposure time and less noise.

AIR GAP TECHNIQUE

- It is used to reduce the scatter radiation.
- Useful in large animal radiography in which use of grid is difficult.
- FFD is increased to 6 feet (decreases the penumbra and magnification) and OFD is increased to 6 inch (less scatter radiation reaches to the cassette).

12

Dark Room and Film Processing

Purpose

1. Storage of unexposed films
2. Loading and unloading of films
3. Processing of exposed films

General

- The dark room must be Clean, Organized and Lightproof.
- Most of the work in the dark room is performed with minimal illumination. Light leaks in the dark room leads to significant film fog.
- Film emulsion is extremely sensitive to heat and humidity. So a dark room should be relatively cool and should have low humidity.
- It is a common fallacy that the walls of the dark room should be dark. The opposite is true. The wall of the dark room should be painted white or cream which produces reflection of the safelight providing a more visible work environment.

Setup of a dark room: The dark room should have two platforms *viz.* dry side and wet side.

1. **Dry side**
   - Here cassettes are loaded and unloaded.
   - Film hangers for each size of film should be hung above the table.
   - Safe light with light bulb of 15 watts or less with brown or dark red filter should be used. The safe light should be 4 feet away from the dry bench.
   - Store the films under the dry table in a film bin to allow easy access for loading cassette.

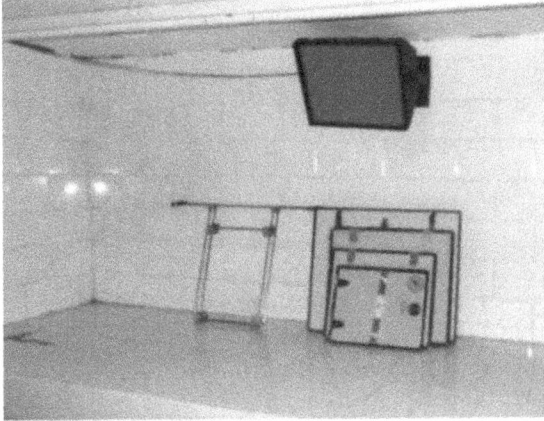

Fig. 12.1a: Dry side of dark room showing safe light, cassettes and film hangers

2. **Wet side**
   - Chemical processing of exposed film is performed here.
   - Consist of 4 tanks containing developer, water, fixer and running water in the central tank.
   - A thermometer is an essential for processing tank because radiographic film is developed for a specified time on the basis of temperature of the chemicals.
   - There should be a film drying area consisting of either a drying rack or a drying cabinet.
   - An illuminator is also recommended to evaluate radiograph on the spot
   - Processing tank: Available in 9, 13 and 22 liters capacity and made up of stainless steel. Developing and fixing tank must always remain covered with lids. Tanks should be washed and cleaned with hypochlorite solution to reduce the risk of fungal, bacterial and algae growth.

FILM PROCESSING

It can be done manually or by automatic machine having 5 basis steps-
   1. Developing
   2. Rinsing or stop bath
   3. Fixing
   4. Washing
   5. Drying

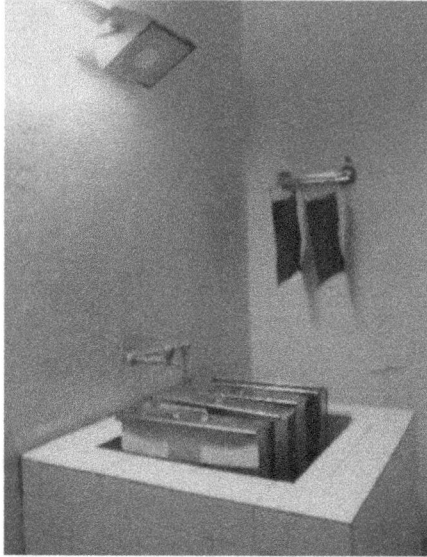

Fig.12.2: Wet side of dark room showing safe light, processing tanks and film drying rack with film hangers

General

- Prepare the developing and fixing solution correctly according to the manufacturer's instructions.
- All the chemical solutions and water should be of the same temperature (Manual tank 68⁰F and automatic processors 95⁰F). Great variation in temperature leads to film reticulation which appears as a mottled density on a finished radiograph. It is caused by irregular expansion and contraction of the film emulsion.

Chemicals used:

1. **Developer:**

   Action: It is used to convert the exposed silver halide crystals to black metallic silver (converts the latent image to visible image on a film). It contains-

   (i) **Reducing agent or developing agent:** *e.g.* Hydroquinone or Phenidone or Metol.

   - Converts the exposed grains of silver halide to black metallic silver.
   - Dark and gray sheds provides contrast to the image.
   - It has little or no effect on the unexposed silver halide crystals.

   (ii) **Accelerators/activator:** *e.g.* Potassium carbonate or sodium carbonate.

- Increases the activity of developer by increasing the pH (9.8-11.4).
- Increase in pH (alkaline medium) soften and swell the emulsion allowing the developing agent to act more effectively on the film.

(iii) Restrainer: *e.g.* Potassium bromide.

- Controls the activity of developing agent to the exposed silver halide crystals in the film. It lessens the chance of film fogging.

(iv) Preservative: *e.g.* Sodium sulphite.

- Prevents the rapid oxidation of alkaline developing agent.

(v) Hardener: *e.g.* ammonium chloride and ammonium sulphide.

- Added to developers in automatic processors because if emulsion swells extensively, it could be damaged by rollers in an automatic processor.

(vi) Solvent: *e.g.* Water to dissolve the chemicals.

> **Developing solution should not be kept for more than 3 months because of its oxidation.**

2. **Fixer**

Action-

(*i*) It clears the unexposed silver halide crystals from film.

(*ii*) It hardens the gelatin coating (fixation) and neutralizes the developer.

Fixer contains-

(*i*) **Clearing or fixing agent:** *e.g.* sodium thiosulphate and Ammonium thiosulphate.

- These agents dissolve and remove the unexposed halide crystals. If these crystals remain, the film darkens and discolours with exposure to light.

(*ii*) **Acidifier:** *e.g.* acetic acid or dilute sulphuric acid.

- Neutralizes the alkaline developer.

(*iii*) **Hardener:** *e.g.* aluminum salts like ammonium chloride and ammonium sulphide.

- Prevents excessive swelling of the gelatinous emulsion.
- Shortens the drying time.

(*iv*) **Preservative:** *e.g.* sodium sulphite.

- Prevents decomposition of fixing agent.

(*v*) **Ant sludge:** *e.g.* boric acid

- Prevents the precipitation of chemical sludge.

(*vi*) **Buffers:** *e.g.*

- Stabilize the acidity even after the addition of alkaline developer by carryover.

(*vii*) **Solvent:** *e.g.* water.

3. **Replenisher:** Strong solution of original developer to replace lost volume during developing process. Use 4 liters of replenisher for fifty 14" x 17" films.

PROCEDURE FOR FILM PROCESSING

1. Loading of the film in a cassette.
2. Unloading of exposed film and loading in hangers.
3. Developing
    - The film loaded on hanger is immersed in the developing tank.
    - Agitate the hanger for 2-3 times to remove any air bubble from the film.
    - Developing time is 4-5 minutes at 20⁰C. If the temperature of developing solution is high, decrease the developing time and vice-versa.
    - Developing for longer period develops fog on the film.
4. Rinsing
    - The film is immersed in the rinse bath/ stop bath containing 128 ml of glacial acetic acid in one liter of water.
    - Rinsing stops overdeveloping and fixer contamination.
    - Rinsing can be done in running water.
5. Fixing the film
    - After draining of excess water, the film is immersed in the fixing solution.
    - Agitate the film 2-3 times to remove any air bubble and new solution comes in contact with film so agitation reduces fixing time.
    - The duration of the fixing process is usually twice the clearing time or until the removal of "milky" appearance i.e. clear or transparent appearance.
6. Washing the film
    - The film should wash for 20-30 minutes.
    - Incomplete film washing leads to film discoloration with passing of time.
    - Non screen film requires longer washing time than screen films.
7. Drying of the film
    - The film should be dried in a dust free area to prevent artifacts.
    - Drying cabinet with heat can be used for drying.

Automatic processing

- It involves the same basic principle as manual processing.
- The film is transported through the processor by a series of rollers similar to conveyer belt in a factory.
- Automatic processors are expensive.
- Automatic processors are used if flow of work is high.
- Advantages over manual method:
  - (i)   Highly standardized procedure.
  - (ii) Dry radiograph is produced within a short period of time.

Recovery of silver from used fixer solution

- Silver can be recovered by metallic replacement or electrolyte recovery or chemical precipitation.
- About 30 gram silver is recovered from 20-25 X-ray film of 14" x 17" size.

13

S. No.	Artifact	Cause	Prevention
1.	Dark radiograph or High density radiograph **Fig. 13.1**	i. High kVp, mA and long exposure time. ii. Short FFD. iii. Use of wrong screen film combination. iv. Overdevelopment due to increased developer temperature, increasing developing time and exposure of the film to visible light. v. Over-measurement of part under examination.	i. Use optimum exposure factors. ii. Use FFD 90-100 cm. iii. Use right combination. iv. Adjust the developing time and temperature accordingly (Manual tank 680°F and automatic processors 950°F) and avoid exposure of film to visible light. v. Part thickness measurement should be proper.
2.	White or light radiograph or low density radiograph **Fig. 13.2**	i. Low kVp, mA and short exposure time. ii. Increased FFD. iii. Underdevelopment due to decreased developer temperature, decreasing developing time and use of exhausted/ diluted developer. iv. X-ray tube failure. v. Decrease in incoming voltage. vi. Accidental loading of 2 films in the same cassette.	i. Use optimum exposure factors. ii. Use FFD 90-100 cm. iii. Adjust the developing time and temperature accordingly and change the exhausted developer After 3 months. iv. Repair or change the X-ray tube v. Adjust the incoming voltage. vi. Load the single film in the cassette at a time.
3.	Lack of radiographic detail **Fig. 13.3**	i. Increased OFD. ii. Blurring of film due to motion of animal, motion of tube or poor screen film contact. iii. Image distortion. iv. Double exposure factors.	i. Proper OFD. ii. Anesthetize the animal, control tube motion and ensure proper screen film contact. iii. X-rays should be directed at the centre of the film. iv. Use proper exposure factors.

Contd...

Contd...

S. No.	Artifact	Cause	Prevention
4.	Arborescent streaks **Fig. 13.4**	i. Improper film handling develops static electricity discharge. ii. Low humidity in the dark room.	i. Handle the film with care. ii. Maintain optimum humidity in the dark room (40-50%).
5.	Crescent shaped mark of black color **Fig. 13.5**	i. Acute bending of film before development.	i. Avoid sharp bending of the film.
6.	**White spots on the film** **Fig. 13.6**	i. Accidental splashing of fixer or water on unprocessed film. ii. Air may be trapped on the film surface during immersion of the film in the developer. iii. Scratches on the emulsion layer. iv. Any radio-opaque material on the body of animal.	i. Work separately on dry and wet area. ii. Agitate the film 2-3 times in the developer. iii. Careful handling of film. iv. Clean the body of animal before radiography
7.	**Black spot on the film** **Fig. 13.7**	i. Accidental splashing of developer on unprocessed film. ii. Leakage of light in the film storage box or defective cassette. iii. Sticking of adjacent films during fixation. iv. During drying of film running of water droplets from hanger over the film.	i. Work separately on dry and wet area. ii. Change/repair the film storage box/cassette. iii. Avoid overcrowding of films in the fixation tank. iv. Remove water droplets if any.
8.	**Fog on the film** **a. Light fog**	i. Slight exposure of film to visible light through cassette, film storage box or defect in the dark room. ii. Safe light: • Use of more wattage bulb. • Crack in the safelight filter. iii. Inspection of film for longer duration during development. iv. Light switch on before fixation.	i. Repair the dark room, film storage box and change the cassette. ii. Use max. 15 watt frosted bulb and intact filter. iii. Don't inspect film for longer duration. iv. Switch on the light after fixation.

Contd...

Contd...

S. No.	Artifact	Cause	Prevention
	b. Chemical fog	i. Development of the film for longer duration. ii. Use of exhausted developer.	i. Develop the film for optimum period. ii. Use new developer solution.
	c. Radiation fog	i. Exposure of film to either primary or secondary radiation/scatter radiation during storage, transport or during radiography.	i. Store the film away from the radiation source. ii. Extra cassettes loaded with films should be kept away during exposure. iii. To check the scatter radiation fog, use grid on parts thicker than 10cm.
	d. Film fog	i. Presence of chemical (Ammonia) fumes in the processing room. ii. High temperature or humidity of the film storage box. iii. Use of expired films. iv. Excessive pressure on the film during storage.	i. The processing room should be free from of any chemicals. ii. Storage conditions should be optimum. iii. Use films before expiry date. iv. Store the films in the upright position.
9.	Blurred radiographic image **Fig.13.8**	i. Motion of the animal or tube during exposure.	i. Restrain/anesthetize the animal and avoid tube motion.
10.	Image distortion **Fig. 13.9**	i. Primary beam off centered. ii. Poor screen film contact. iii. Deviation of angle of central beam to the cassette.	i. Center the primary beam. ii. Clean the intensifying screen by screen cleansers only. iii. The central beam should be perpendicular to the cassette.
11.	Grid lines	i. Selection of wrong FFD. ii. Off centered grid. iii. Central beam is not at right angle to the grid. iv. Use of reverse grid.	i. Use FFD as recommended. ii. Center the primary beam over the center of grid. iii. Central beam should be at right angle to the cassette. iv. Use grid in proper position.
12.	Yellow radiograph **Fig. 13.10**	i. Use of exhausted fixer. ii. Fixing for short duration.	i. Discard the exhausted fixer. ii. Fix the film properly.

Contd...

Contd...

S. No.	Artifact	Cause	Prevention
13.	Frosty area on the film **Fig. 13.11**	i. Incomplete final washing of the film.	i. Give sufficient final washing to the film.
14.	Frilling of gelatin **Fig. 13.12**	i. High temperature of the processing solution.	i. Maintain the temperature within limits.
15.	Finger marks **Fig. 13.13**	i. If fingers are contaminated with- • Fixer: white marks • Developer: black marks ii. Greasy or wet hands.	i. Hands should be cleaned and dried before handling of films.
16.	Reticulation	i. Large difference in temperature of developer and fixer. ii. Too old fixer solution.	i. Temperature of all the solutions and water should be kept same. ii. Change the fixer.
17.	White horizontal area on top of film **Fig. 13.14**	i. When the film is not completely immersed in the developer.	i. The film should be completely immersed in the developer.
18.	Dark horizontal area on the top of film **Fig. 13.15**	i. When the film is not completely immersed in the fixer.	i. The film should be completely immersed in the fixer.
19.		i. Film is not exposed. ii. Film away from radiation beam exposure. iii. If exposed film placed in the fixer before the developer.	i. Check the electric supply and X-ray machine. ii. Place the cassette in the center of the primary beam. iii. First develop the film.

Contd...

Contd...

S. No.	Artifact	Cause	Prevention
20.	Blank film/ Clear film	i. Film agitation not performed. ii. Improper rinsing of film. iii. Use of dirty water for washing of film. iv. Use of dirty film hanger.	i. Agitate the film 2-3 times. ii. Rinse the film properly. iii. Use fresh water for washing. iv. Use clean film hanger.
21.	Streak on the film	i. Damaged roller of automatic processor cause scratches.	i. Repair/change the roller.
22.	White lines in the length of film Uneven development	i. Use of unstirred chemical that settle to the bottom of tank. ii. Repeatedly viewing of film during developing process.	i. Stir the chemical before immersion of the film. ii. Don't view the film repeatedly during developing process.

14

Contrast Radiography

Contrast radiography can be defined as deliberate alteration of the radio-density of the tissue/organ or its surrounding structures to enhance the demarcation and visualization of these organs on the finished radiograph. The agents used for this purpose are called contrast agents/media.

OBJECTIVE

- To differentiate an organ from the surrounding tissue.
- To determine the size, shape, position, location and function of an organ.
- To determine the abnormality on serosal or mucosal surface of hollow organs.
- A contrast study should never be used in place of survey radiographs, Survey radiograph should always be taken before administering contrast media.

In contrast radiography the organ appears-

1. **Radiopaque or white:** when a positive contrast agent is used.
2. **Radiolucent or black:** when a negative contrast agent is used.

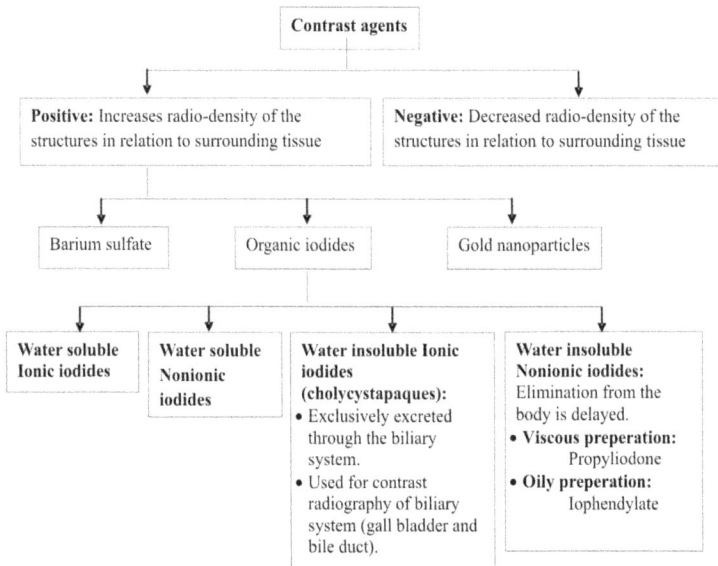

High osmolar contrast media (HOCM)	Low osmolar contrast media (LOCM)
• More toxic • All ionic contrast media	• Less toxic • Most non-ionic contrast media· • First LOCM (Metrizamide) was produced by Nyegaard

Adverse reaction

1. Anaphylactoid reactions: Urticaria itching, bronchospasm, facial and laryngeal oedema
2. Contrast induced nephropathy

POSITIVE CONTRAST MEDIA: These agents are inert, have high atomic number (high density) and absorbs more X-rays making a white area on the radiograph.

1. **Barium sulfate**

   • It is marketed in the form of powders, suspension and paste.

   • Mostly used for positive contrast radiography of gastrointestinal tract.

   • It is insoluble and unaffected by gastric secretions.

   • It provides good mucosal detail on radiograph.

Disadvantages

   • It should never be used in esophageal and gastrointestinal perforation because it cannot be eliminated from the body and causes granulomatous reactions.

   • It takes long time (3 or more hours) to travel from the stomach to the colon.

   • If accidently goes to lungs, it may be fatal. So it should be administered carefully to prevent the patient from aspirating.

   • It may aggravate the condition in cases of obstructed bowel by causing further impaction.

2. **Organic iodides**

   • These are water soluble organic iodides that can be administered intravenously, orally in to a hollow visceral organ or into the subarachnoid space.

   • The iodides are excreted by kidneys.

   • Organic iodides are well tolerated by the body and provide excellent contrast.

A. Ionic iodides

- These iodides should never be used in place of barium sulfate and used only when perforations are suspected. Iodides transit rapidly through the G.I. system (within 45-60 minutes).
- Ionic iodides are hypertonic solution that draws fluid into the bowel lumen. So it should be used in diluted form in dehydrated animals (because fluid loss into the lumen causes further hypovolemia).

Indications

- Urinary bladder or fistulous tract.
- Sodium diatrizoate for excretory urography.

Contraindications

- Should not be used for myelography because they are irritating to the brain and spinal cord.
- Water soluble iodine preparations are contraindicated in dehydrated animal because of their hypertonic properties

Side effects: Nausea, vomiting and decreased blood pressure can occur after rapid administration of large intravenous bolus of contrast media.

B. Non-ionic iodides

- Used for myelography and can be administered intravenously.
- Ten times more expensive than ionic media.
- Example: Iohexol, Iopamidol and Iodixanol.

NEGATIVE CONTRAST MEDIA: An ideal negative contrast media should be inert, quickly dissolved in the body fluids and quickly eliminated from the body. Elimination of CO_2 is very quick because solubility of CO_2 in water is very high.

- Gases of low atomic number are used.
- On radiograph there is a radiolucent area.
- Don't overinflate the hollow organs like urinary bladder with negative contrast media. Over-inflation leads to air embolism or rupture of the organ.
- Example: Room air and Carbon dioxide (CO_2), O_2 and N_2.

Indications: Arthrography, Fasciagraphy, Pneumoperitoneography and Pneumocystography.

Table 14.1: Commonly used contrast agents

S.No.	Trade name	Generic name	Indications
Barium sulfate preperations			
1.	Novopaque	Barium sulfate suspension	Oesophagography,
2.	Baritop-100	Barium sulfate suspension	reticulography, gastrography,
3.	Baritop-G	Barium sulfate powder	barium meal for GIT, barium enema
4.	Microtrast	Barium sulfate paste	
Water soluble Ionic iodides			
1.	Hypaque sodiums	Sodium diatrizoate	Angiography,
2.	Hypaque meglumine	Meglumine diatrizoate	Sialography,
3.	Renografin 60	Sodium diatrizoate (8%) and	Phlebography,
		Meglumine diatrizoate (52%)	Osteomedulography,
4.	Renografin 76	Sodium diatrizoate (10%) and	Arthrohraphy,
		Meglumine diatrizoate (66%)	Cystography,
5.	Conray 60, Conray 280	Meglumine iothalamate	Urethrography,
6.	Conray 325, Conray 420	Sodiume iothalamate	(IVP),
7.	Uromiro 380, Uromiro 420	Meglumine iodamide	Intravenous pyelography
8.	Uromiro 300	Sodium iodamide	Dacrocystorhinography,
9.	Isopaque 350,	Sodium Metrizoate	Double contrast peritoneography,
	Isopaque 440		Vertebral venography
10.	Hexabrix	Ioxaglate	
Water soluble Non-ionic iodides			
1.	Amipaque	Metrizamide	Myelography and all
2.	Isovue-M 200, Niopam	Iopamidol	indications of Water
	200, 300 and 370		soluble Ionic iodides
3.	Omnipaque 180, 240	Iohexol	
4.	Visipaque 270, 320	Iodixanol	
Viscous and oily preparations			
1.	Dionosil-oily	Propylidone	Lymphangiography,
2.	Lipiodol-ultra	Iodized poppy seed oil	Dacrocystorhinography,
3.	Myodil	Iophendylate	Hysterosalpingography,
			Bronchography,
			Sialography
Cholecystopaques			
1.	Telepaque	**Oral:**	Cholecyctography (contrast
		Iopanoic acid, Sodium	radiography of gall bladder and bile
		iopodate, Iocitamic acid	duct)
2.	ChlografinBiligrafin	**Intravenous:**	
		Calcium iopodate, Meglumine	
		salt of Iodipamide, Iodoxamate,	
		Ioglycamate, Iotroxate and	
		Iodipamide	

CONTRAST STUDIES OF DIFFERENT BODY SYSTEMS

Orbital angiography: Contrast material is injected through the infraorbital artery.

Indications

- Diagnosis of neoplasm
- Diagnosis of vascular abnormalities

Materials required

- Vein infusion set and disposable syringe.
- Heparinized physiological saline.
- Iodine based water soluble contrast media e.g. Hypaque (50%).
- General surgical pack for isolation of artery.

Procedure

- Anesthetize the patient and prepare the site (infraorbital foramen region) aseptically.
- After skin incision expose the infraorbital artery. Cannulate the artery in proximal direction with 21 gauze vein infusion set and needle is tied in position by preplaced ligatures.
- Place the side of examination near the film and infuse 5-10 ml of contrast medium in the artery.
- Take a radiograph near the end of the injection.

Interpretation

- Displacement of vessels and abnormality in vascularization due to masses in the orbital region.

Complication

- May cause orbital pain, ocular hemorrhage and transient or permanent blindness.

Sialography: Contrast radiographic examination of salivary glands and its ducts.

Indications

- Salivary mucocele or fluid containing swelling.
- Recurrent swelling at cheek, parotid sialography should be done.
- For confirmatory diagnosis of sialolithiasis, sialoangiectasis, sialadenitis, abscess and tumor.

Contrast agent and doses: 60% suspension of propylidone in peanut oil (Dionosil oily) @ 0.1-0.3 ml for retrograde injection in to the salivary duct.

Procedure

1. **Parotid gland and duct**
   - Parotid duct opens opposite to the upper 4th (carnassials) tooth on the labial surface.
   - A 20 or 22 gauze needle is inserted into the duct opening and pushed caudally.

2. **Zygomatic gland and duct**
   - The major duct of zygomatic salivary gland opens into the mouth cavity at 1 cm caudal and slightly dorsal to the parotid duct opening.
   - A 25-26 gauze needle is inserted into the duct opening and pushed caudally.

3. **Mandibular and sublingual duct**
   - The mandibular salivary duct opens by a slit like opening on the lateral surface of the lingual caruncles.
   - The sublingual duct opens 1-2mm caudal to the mandibular duct opening and appears as red spot (in most dogs).
   - Both ducts joins and opens in mouth by a common opening (in some dogs).
   - A 25-26 gauze needle is inserted into the duct opening and pushed caudally.

After cannulation, inject the contrast media and make a lateral and ventro-dorsal view (lateral view is more informative).

Interpretation

1. *Parotid gland and duct:* The parotid gland is an irregular lobulated structure that lies latero-ventral to the external auditory meatus.

2. *Zygomatic gland and duct:* It is small single lobed gland which lies ventral to the rostral end of the zygomatic arch.

3. *Mandibular gland and duct:* It is a single lobed gland. Lateral view is superimposed by hyoid bone. This should not be mistaken for extravasation of the contrast medium from the duct or gland.

4. *Sublingual gland and duct:* The main part of the gland is located at the level of the 2nd molar tooth and courses caudoventrally.

Dacrocystorhinography: Contrast radiographic examination of nasolacrimal duct.

Indications

- Chronic or intractable conjunctivitis
- Dacrocystitis
- Neoplasms of the lacrimal duct or periductal tissue

Procedure

- After general anesthesia, place the patient in lateral recumbency with diseased side up.
- Superior punctum lacrimale is cannulated with 20 gauze beveled polyethylene catheter.
- Apply fixation forceps to the inferior punctum.
- Flush the nasolacrimal duct with normal saline
- Water soluble iodine based contrast medium (50-60%) is injected through the catheter until the contrast agent emitted through the nostril.
- After injection, turn the animal and place the side to be examined closest to the film.
- Make a lateral radiograph and open mouth view.

Pneumopericardiography: It is the negative contrast radiographic examination of pericardial sac.

Indications

- When etiology of pericardial fluid cannot be determined from other methods.

Procedure

- Rapid digitalization, administration of diuretics and narcotic sedatives.
- Prepare the left ventral thorax (left 4th – 6th intercostals space near the junction of lower and middle third of the ribs) aseptically.
- Place the animal in right lateral recumbency and infiltrate the local anesthetic at the site.
- The needle is inserted through the thoracic wall in medio-dorsal direction at 45^0.
- The catheter and wire guide are inserted into the pericardial sac through the needle. Thereafter the needle is retracted from the sac.
- A 3 way stopcock and syringe are then attached to the catheter.
- Fluid is removed and air is injected into the pericardial sac.
- Make dorso-ventral, right and left lateral and standing lateral view.
- After procedure, remove the catheter and antibiotic therapy is started.

Contraindications

- Congestive heart failure
- Pulmonary edema

Complications

- Rupture of coronary vessels - severe hemorrhage.
- Ventricular arrhythmia.

Angiocardiography: Sequential production of radiographs during the injection and circulation of the contrast agent through the heart and blood vessels.

Indications

1. Selective angiocardiography: By delivering the contrast agent near the suspected lesion through catheter.
   - If routine method have failed.
   - Congestive heart defects: e.g. ventricular septal defects, valvular insufficiency etc.
2. Non- selective angiocardiography:
   - Differentiating cardiomyopathies in cat-
     - Congestive cardiomyopathy: Dilated left ventricular cavity.
     - Hypertrophic cardiomyopathy: Constricted left ventricular cavity.
   - Pericardial diseases: Increased distance between endocardium and pericardium.
   - Pulmonic stenosis
   - Right to left shunts
   - Right atrial and ventricular masses.

Procedure

- Anesthetize the patient and prepare the site over the carotid artery or femoral artery.
- The catheter is passed along the artery up to the desired position and flush with anticoagulant.
- Inject the contrast agent and take the radiograph.

Pneumoperitoneography: It is the negative contrast study of the abdominal cavity after injection of air/gas into the peritoneal space for increasing the subject contrast.

Indications

- Routine radiography failed to demonstrate the abdominal organs/masses.
- Accumulation of excessive peritoneal fluid.
- Diaphragmatic hernia.

Procedure

- Patient is fasted for 12 hr.
- After inducing anesthesia, mid-ventral abdomen is prepared aseptically.
- Place the catheter-needle assembly into the abdomen approximately 1 cm lateral to the umbilicus.
- Withdraw the needle and attach a three way valve and syringe.
- Peritoneal fluid should be drained before pneumoperitoneography.
- Inject the gas (200-1000 cc. or until moderate distension of the abdomen) and make VD and lateral views after completion of injection of air.
- Dorso-ventral view is required for viewing the structures in dorsal abdomen.
- Erect, inverted and standing radiographs can be taken with horizontal X-ray beam.
- If diaphragm is ruptured, pneumothorax will develop. So a positive contrast peritoneogram is done to diagnose diaphragmatic hernia. Inject sodium/meglumine diatrizoate @ 1.5 ml / kg body weight and rotate the patient on its longitudinal axis. Make a lateral radiograph.

Cholecystography: Contrast radiographic study of gall bladder and bile ducts.

Indications

- Diseases of gall bladder or bile duct *e.g.* cholecystitis, cholelithiasis, cholengiectasis or neoplasm of gall bladder and bile ducts.

Procedure

- The contrast agent (meglumine iodipamide @ 0.2 ml/kg) is injected in to the cephalic vein slowly (up to the 3 minutes because rapid injection causes retching and discomfort).
- Make a radiograph of cranial abdomen at 30, 60 and 90 minutes after injection.
- If post emptying study of the gall bladder is required, fed a fatty diet and make a radiograph in about 15 minutes.

Contraindications

- Hypersensitivity to organic iodides.
- Salts of iodipamide are contraindicated in severe renal and liver disease and hyperthyroidism.

Interpretation

- Alteration in margination content is a sign of disease condition.

CONTRAST RADIOGRAPHIC STUDY OF ELEMENTARY TRACT

Fluoroscopy: Image intensified fluoroscopy can evaluate the peristalsis, transit time, mucosal integrity and luminal shape, size and content.

Routine radiography: Peristalsis cannot be evaluated.

Esophagogram/ barium swallow: Radiographic investigation of the hypopharynx and esophagus.

Indications

- Esophageal foreign bodies and neoplasms
- Reflux esophagitis
- Esophageal diverticula and deviation of the esophagus.
- Persistant gagging or vomiting and regurgitation of undigested food.
- Megaesophagus and achalagia

Contrast material and doses

- Thick suspension or esophageal paste of barium sulfate is preferred for esophagus because thick suspension coats the esophageal mucosa better than the watery media.
- For upper G.I. series, liquid suspension is preferred.
- Dose: 2-6ml of contrast agent/ kg body weight.

Procedure

- Off fed the animal for 12 hour.
- Administer the barium sulfate orally with a syringe.
- Make a radiograph just after the administration of the contrast agent.

Complications

- Aspiration of contrast agent.
- Leakage and accumulation of contrast agent through the esophageal perforation.

Contraindications

- Esophageal rupture and perforation.
- Bronchoesophageal fistulae.
- Inability to swallow may lead to aspiration pneumonia.

Interpretation

- Barium sulfate coats between the folds of esophageal mucosa and appears as parallel lines of nearly uniform width.

- In cats, in addition to the longitudinal folds, the caudal 3rd of esophagus is transversely striated. This produces a herring-bone or burlap pattern.

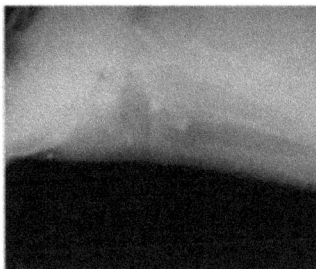

Fig. 14.1: Esophageal diverticulum in a dog

Gastrography: contrast radiography of the stomach.

Indications

- Intramural and intraluminal gastric masses.
- Radiolucent gastric foreign bodies.

Contraindications

- In the presence of ingesta or food.
- Diarrhoea
- Double contrast radiography is preceded by parentral administration

S. No.	Technique	Negative contrast gastrography or pneumogastro-graphy	Positive contrast gastrography	Double contrast gastrography
1.	Patient preparation	Off fed the animal for 12 hour	Off fed the animal for 12 hour	- Off fed the animal for 12 hour - Stomach is paralysed (hypomotility) by intravenous Glucagon (Dose should not exceed 1.0mg) or Propanthalene @ 7.5-30 mg orally
2.	Materials required and dose rate	-Effervescent agents for gastric distension. OR -Highly carbonated beverage (30-60 ml orally). OR	- Barium sulfate suspension @ 6-12ml/kg body weight and concentration should not exceed 15% w/w. Micropulverized BaSO$_4$ is preferred over plain BaSO$_4$. Because later flocculate in the lumen	- Barium sulfate Micropulverized: 3ml/kg for animals upto 8 kg 2ml/kg for animals upto 8-40 kg 1.5ml/kg for

Contd...

Contd...

S. No.	Technique	Negative contrast gastrography or pneumogastrography	Positive contrast gastrography	Double contrast gastrography
		-Room air, stomach tube, three way valve, mouth gag and syringe. -Room air @ 6-12ml/kg body weight through the stomach tube	and will not coat the mucosal surface and its passage through the small intestine is slower as compared to the former one - Aqueous organic iodide solutions are irritative and induce hypermotility. In cat pylorospasm and vomiting occurs due to pyloric irritation and hyperosmolaric increase in the gastric volume. -Dose of the organic iodides should not be > 10% w/v - Mural and luminal masses may be hidden if the material is too dense.	animals > 40 Kg Followed by gastric insufflations with room air @ 20 ml/kg
3.	Procedure	-Administration of room air OR effervescent agents OR Highly carbonated beverage via the stomach tube. -Remove the stomach tube. -Immediately take radiograph of cranial abdomen in VD, DV and left or right lateral position	-Administer micropulvarized $BaSo_4$ OR Diatrizoate solution via stomach tube and roll the animal on its spinal axis. -Immediately take radiograph of cranial abdomen in VD, DV and left or right lateral position. **Fig. 14.2:** Positive contrast gastrography in a dog	-Induce gastric hypomotility or paralysis. -Administer $BaSO_4$ -Insufflate the stomach by room air. -Remove the tube and roll the patient to coat the gastric mucosa. -Immediately take radiograph of cranial abdomen in VD, DV and left or right lateral position.

of glucagon. So contraindicated in diabetes mellitus or suspected pheochromocytoma.

Upper gastrointestinal series (Barium series, small bowel series): A radiologic study in which barium sulfate is swallowed, providing an image of the upper gastrointestinal tract including the esophagus, stomach, and duodenum.

Indications

- Recurrent and nonresponsive vomiting.
- Recurrent diarrhea and hematemesis.

- Intestinal obstruction.
- Localization of abdominal masses and neoplasm.
- Gastrointestinal radiolucent foreign bodies.
- Confirmation of hernia e.g. diaphragmatic and perineal hernia.

Contraindications

- Perforation of gastrointestinal tract because it causes granuloma formation in the peritoneal cavity. In perforation / rupture water soluble iodine preparations should be used.
- Obstructive ileus or small bowel atony.
- Oral diatrizoates are contraindicated in debility because of their hyperosmolality and subsequent dehydrating effect.

Patient preparation

- Off fed the animal for 24 hr and enema by salt solution.
- Soapy enemas are not generally used due to its surfactant property. Surfactants coalescence the gas which does not expelled the solid fecal material.
- Sedatives or tranquilizers should not be given before undergoing barium series because these agents modify the gastrointestinal motility. So small intestine transit time is modified.
- Anticholinergics should not be given within 24 hr of radiographic examination.

Contrast agents:

- Micropulverized preparation of $BaSO_4$ is best for barium series.
- In perforation organic iodides are used.
- For quick radiographic study organic iodides are used because these agents pass very rapidly through the small intestine.
- 15-20% suspension of $BaSO_4$ @ 6-12 ml/kg in dogs.
- The water soluble organic iodides @ 2 ml/kg body weight. Don't dilute these iodides since these solutions lose density as they pass through the G.I. tract.

Procedure

- Slowly administer the calculated dose of contrast agent through the stomach tube or by buccal pouch.
- The organic iodides are bitter that's why these agents should be administered through the stomach tube only.
- Make lateral and VD radiographs of abdomen.
- Immediately after administration.

- 15 minutes, 30 minutes, 1, 2, 3hr until the contrast material **(BaSO$_4$)** reaches the colon.
- 5, 15, 30 and 60 minutes and then every half hour until the contrast material **(organic iodides)** reach the colon.

Complications

- Aspiration of large amount of BaSO$_4$ is fatal.
- Aspiration of vomited barium (even in small amount) is more dangerous due to presence of gastric acid which affects the upper airway epithelium.
- Peritoneal granuloma if G.I. lumen is perforated.
- Organic iodides are harmful to dehydrated and debilitated patient because these agents cause further dehydration because of hyperosmolality.
- In cats organic iodides cause pylorospasm due to its irritant property which causes gastric retention and vomition.

Interpretation

S. No.	Species	Contrast agent used	Small intestine transit time
1.	Dog	Micropulvarized BaSO$_4$	2-3 hr
		BaSO$_4$ (USP)	3-4 hr
		Organic iodides	45 minutes-1 hr
2.	Cat	Micropulvarized BaSO$_4$	30-60 minutes
		BaSO$_4$ (USP)	30-60 minutes

- A smooth halo is seen surrounding the luminal column of contrast material except in ileum where the wall is very smooth. The halo is caused by presence of contrast agent in the crypts between the intestinal villi.
- The halo effect is more with organic iodides as compared to micropulvarized BaSO$_4$.
- The mucosal borders are uniform except in duodenum, where the crater shaped pseudoulcers are seen along the antimesentric border.
- In cats, in upper G.I. series, the normal duodenal pattern is like "string of pearls".

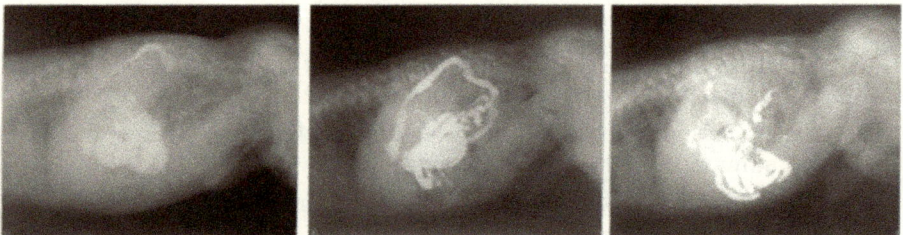

Fig. 14.3: Barium series in a dog

Barium enema, lower bowel series, and double contrast radiography of the colon: Contrast radiographic study of colon (Position and contour).

Indications

- Recurrent bloody diarrhea/large bowel diarrhoea.
- Low volume, high frequency stool with or without blood.
- Rectal tenesmus.
- Intussusception at the iliocolonic junction.
- Neoplasms of colon and rectum.
- Colitis.

Contraindications

- Colonic obstruction.
- Perforation/ rupture of colon.
- Recent biopsy.
- Lesion limited to rectum or terminal colon.
- Patient undergoing proctoscopy within 12 hr before study and cleansing enema within 4 hr before study.

Procedure

- Patient preparation:
  - Off fed the animal for at least 24 hour.
  - Give mild cathartics and enema should be given night before.
  - General anesthesia should be given because inflation of the bulb of cuffed catheter causes discomfort.
- Place the animal in right lateral recumbency.
- Place the cuffed rectal catheter or enema catheter or foley catheter in rectum and inflate the cuff to prevent leakage of contrast agent.
- Infuse the contrast agent (15-20% suspension of micropulverized barium sulfate @ 20-30 ml/kg body weight) through catheter in to the colon and continue until proper distension of colon is achieved.
- Make a VD and lateral radiograph of the abdomen.
- For double contrast study-
  - Remove the contrast material by the same catheter.
  - Fill the colon with a volume of air equal to the amount of contrast material recovered.
  - Make a VD and lateral radiograph of the abdomen.
  - Deflate the cuff and catheter is removed.

Complications

- Rupture of colon if over-distension occurs.

Interpretation

- Colon forms like "question mark" or a "shepherd's crook" in VD view

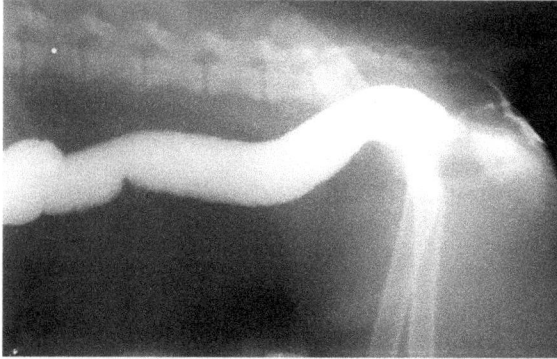

Fig. 14.4: Barium series in a dog

CONTRAST RADIOGRAPHY OF THE URINARY SYSTEM

Intravenous pyelography (IVP) / excretory urography / Intravenous urography: IVP gives information about the renal function and structure of the kidneys and ureters.

Indications

- To demonstrate the kidney's capacity to concentrate and excrete circulating organic iodinated contrast media.
- Palpable abnormality of kidney and ureters.
- Suspected renal masses and prostatic masses.
- Persistent hematuria and abnormal urine.
- Renal/Ureteral calculi.
- Hydronephrosis.
- Ectopic ureter, kidney and ureterocele.
- Exstrophy of the bladder.
- Congenital urinary tract abnormalities.
- Traumatic rupture of the ureter.
- Can be used to evaluate urinary bladder. But it cannot be completely evaluated from the IVP alone because proper distension of urinary bladder cannot be achieved.

Contraindications

- Severe dehydration and debilitation.

Procedure

- Patient preparation:
  - Make a survey radiograph
  - Off fed the animal for 24 hr.
  - Enema by isotonic saline solution
  - Empty the urinary bladder
- Materials required
  - Intravenous infusion device should be used because contrast materials results in severe sloughing of the tissues.
  - Triiodinated contrast agents are used. Diatrizoate salts are safe in uremia.
  - Compression device- It presses the urinary bladder against the spine and occludes the terminal ureters causing stasis and accumulation of opacified urine in the renal pelvis and proximal ureters.

Low volume rapid infusion with abdominal compression	Low volume rapid infusion without abdominal compression	Low volume slow infusion without abdominal compression	High volume drip infusion with abdominal compression
1. Catheterize the peripheral vein	1. Catheterize the peripheral vein.	1. Catheterize the peripheral vein.	1. Catheterize the peripheral vein.
2. Apply the compression device just cranial to the pubis	2. Place a loaded cassette and position the animal over it in dorsal recumbency.	2. Infuse the contrast agent @ 425 mg Iodine/ kg Intravenously over a period of 3 minutes.	2. Make a mixture of 1200 mg Iodine/ kg body weight with equal volume of 5% Dextrose and water.
3. Rapidly infuse the contrast agent @ 425 mg Iodine/ kg Intravenously.	3. Rapidly infuse the contrast agent @ 850 mg Iodine/ kg Intravenously.	3. Make VD and lateral radiograph of abdomen immediate and at 3, 5, 10 and 15 minutes after injection.	3. Infuse the mixture by drip infusion over a 10 minute period.
4. Make VD and lateral radiograph of abdomen immediate and at 1,3, 5 and 10 minutes.	4. A VD radiograph as made after 10 seconds of injection.	4. VD view of pelvis can be made to visualize terminal ureters.	4. Increase the rate of infusion near the end and make the VD and lateral radiograph of abdomen at the end of infusion.
5. Release the abdominal compression prior to 10 minute radiograph.	5. Make VD and lateral radiograph of abdomen immediate and at 1, 5 and 15 minutes.	5. Visualization of the collecting system is poor	5. Apply a compression band to the caudal abdomen immediately.
6. Terminal ureters or ectopic ureters can be best visualized by making oblique VD radiograph after releasing the compression.	6. Ureteral opacification is less pronounced than with abdominal compression		6. Make a radiograph at 10 and 20 minutes.
			7. Remove the compression band

Complications

- Rapid infusion of the contrast agent leads to vomiting. This effect is more common when sodium salt is used.
- Severe tissue sloughing if contrast agent is deposited perivascularly.
- Anaphylactic reaction.
- Pancytopenia if diatrizoates are used
- Abdominal compression is hazardous in the presence of an abdominal mass.

Interpretation

- Normal position of the kidneys:
  - Right kidney-T_{13} to L_3
  - Left kidney- L_2 to L_5
- Normal kidney length:
  - In canine: about $2^1/_2$ time the adjacent vertebral bodies.
  - In feline: about 2 time the adjacent vertebral bodies.
- The radiographic ratio of kidney length to the length of L_2
  - In canine: about 2.5:1 to 3.5:1
  - In feline: about 2.4:1 to 3.0:1

Cystography/ Pneumocystography: Contrast radiography of the urinary bladder.

Indications

- Straining to urinate with or without hematuria/ vesicle tenesmus.
- Radiolucent urinary calculi.
- Localization of urinary bladder in hernia. (Perineal hernia).
- Caudal abdominal or pelvic masses.
- Enlargement of the prostate.
- Ruptured urinary bladder-only cystography is indicated.
- Congenital anomalies-Pervious urachus.

Contraindications

- Massive bladder enlargement due to atony of the bladder
- Pneumocystograph sometimes may lead to air embolism specially when the mucosa of the urinary bladder is damaged

Procedure

- Preparation of the patient:
  - Off fed the animal for 24 hr.
  - Enema (Colon should be free from fecal material).

- Materials required:
  - Catheter and a 3 way stopcock.
  - Positive contrast media (Diatrizoates).
  - Room air for pneumocystography.
  - Dose: 5-10% iodine w/v of contrast material @ 6-12 ml/Kg body weight.
- Make a survey radiograph.
- Induce general anesthesia.
- Catheterize the bladder and remove the urine from the bladder.
  - Sodium iodide solution should not be used because it causes irritation of the bladder mucosa.
  - In ruptured bladder it will produce a transient but severe peritonitis.

Positive contrast cystogram	Pneumocystogram	Double contrast cystogram
1. Inject the positive contrast agent slowly while palpating the bladder. 2. If resistance is feeling, stop the injection 3. Make VD and lateral radiograph of caudal abdomen.	1. Infuse the air slowly while palpating the bladder 2. Make VD and lateral radiograph.	1. Inject the positive contrast agent slowly while palpating the bladder. 2. If resistance is feeling, stop the injection 3. Make VD and lateral radiograph of caudal abdomen. 4. Remove the contrast agent and replace it with equal volume of air. 5. Make VD and lateral radiograph

Interpretation

- Mucosal surface is best interpreted with positive contrast agent.
- Bladder wall is best interpreted with negative contrast agent.

Complications

- Overdistension and rupture of urinary bladder.
- Urethral trauma during catheterization.
- Cystitis.
- Fatal air embolism.

Urethrography: Contrast radiographic study of urethra.

- It can be divided into prostatic urethrography and prostatic urethrography.
- It is difficult in female because urethra is relatively short and wide.

Indications

- Dysurea, tenesmus and/or hematurea.
- Indications for retrograde urethrography:

- When urethral catheterization is difficult e.g. penile or extrapelvic urethral disease.
- Trauma to urethra, prostate and pelvis.
- Fracture of os penis.
- Indications for prostatic urethrography:
  - In prostatic diseases.
  - Urinary incontinence, deviation of urethra.
  - Obstruction/calculus.

Contraindications

- High pressure should be avoided during retrograde urethral filling.
- Severe urethral hemorrhage.

Procedure

- Give enema for cleaning the terminal colon and rectum.
- Anesthetize the patient and make a survey radiograph.

Retrograde urethrography	Voiding Urethrography
1. The foley catheter lumen is filled with 2% Lidocaine (in male) and contrast media (in female) before catheterization in order to eliminate the air bubble. 2. Apply lidocaine gel and catheter is introduced in to the urethra. 3. Position the animal in lateral recumbency 4. Inject local anesthetic in to the penile urethra through the foley catheter. 5. Inflate the cuff. 6. Slowly inject 5-10 ml contrast agent (Diatrizoates). While injection slowly withdraws the catheter and when the catheter tip reaches distal urethra, inject a bolus of contrast agent. 7. Make a radiograph as the injection is terminated. 8. Lateral projections are important in male dogs.	1. Place the animal in lateral recumbency. 2. Inflate the bladder with diluted positive contrast agent. 3. Cassette is placed and external pressure is applied to the bladder by the wooden spoons. 4. Make a radiograph when contrast agent is flowing from the urethra.

Complications

- Air bubble misinterprets calculi.
- Urethral trauma due to over distension of cuff of foley catheter.

NEURORADIOGRAPHY

Myelography: Contrast radiographic evaluation of the spinal cord after injecting positive contrast agent in to the subarachnoid space.

Indications

- Spinal compression resulting from a disc protrusion.
- Malformation of the cervical spinal canal.
- To evaluate the localized myelopathies
- For the diagnosis of meningeal tumors.
- To determine the nature exact location and extent of lesion prior to surgery.

Contraindications

- Diffuse myelopathy and meningitis.

Contrast agent and doses

- Metrizamide-Non ionic biologically inert preparation @ 0.3-0.5 ml/kg body weight.

Procedure

1. Cisternal technique:
   - Anesthetize the dog and place in sternal / lateral recumbent position.
   - Aseptic preparation of the site over the occipital protuberance and atlas.
   - Head is held firmly at right angle (90°) to the cervical spine.
   - A 3 inch, 22 gauze spinal needle with stylet is inserted in the dorsal midline with the bevel of the needle directed posteriorly midway between the occipital protuberance and most palpable part of the wing of the atlas.

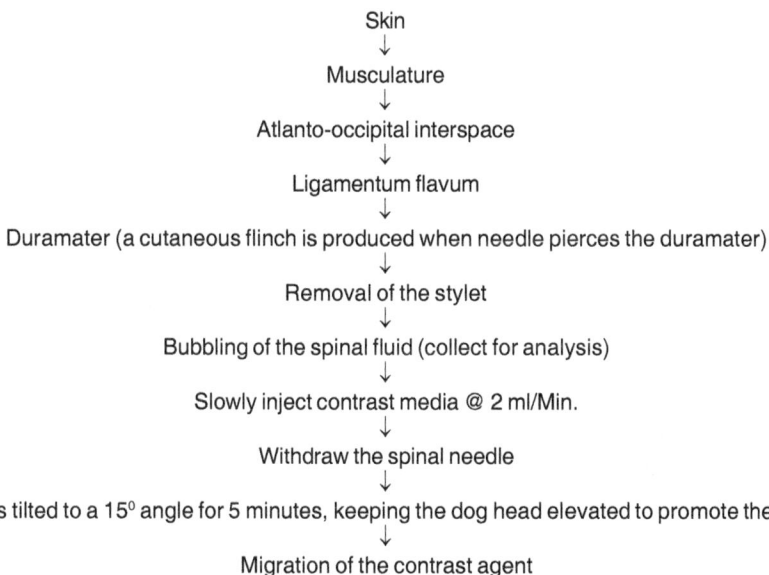

Skin
↓
Musculature
↓
Atlanto-occipital interspace
↓
Ligamentum flavum
↓
Duramater (a cutaneous flinch is produced when needle pierces the duramater)
↓
Removal of the stylet
↓
Bubbling of the spinal fluid (collect for analysis)
↓
Slowly inject contrast media @ 2 ml/Min.
↓
Withdraw the spinal needle
↓
Table is tilted to a 15° angle for 5 minutes, keeping the dog head elevated to promote the caudal
↓
Migration of the contrast agent

- Take lateral and ventro-dorsal radiograph.
2. Lumbar tap technique:
    - Anesthetize the dog and place in flexed sternal / lateral recumbent position.
    - Aseptic preparation of the site.
    - In lateral recumbent position, the hinds should be brought forward between the front limbs in order to open the interarcuate spaces.
    - A 3 inch 22 gauze spinal needle is inserted between L_4-L_5 (in large dogs) or L_5-L_6 interspace (in small dogs and cats) or posterior to the lesion.
    - Insert the needle in the dorsal midline just lateral to the dorsal spinous process of L_6 at 50-60⁰ angles in cranio-ventral direction.
    - Inject the contrast agent @ 0.45 ml/Kg (for cervical study) or 0.3 ml/kg body weight (For thoracolumbar study) in either dorsal or ventral (transmedullary lumbar myelography) subarachnoid space.
    - When the contrast agent is injected in the ventral subarachnoid space, the technique is called as transmedullary lumbar myelography. The complication is myelomalacia.
    - Make a lateral and VD radiograph.

Complications

- Seizures.
- Exacerbation/ aggravation of the clinical signs.
- Apnea may occur during injection of contrast agent.
- Puncture of venous sinuses and hematoma formation.

Interpretation

- Extradural masses: Hematoma, protruded disc and tumor.
- Intradural-extramedulary masses: Tumors compress and deviate the spinal cord.
- Intramedullary masses: e.g. tumor. Spinal cord enlarges in all direction causing narrowing/obliterating the subarachnoid space.
- Tumors and wobbler lesion causes hour glass appearance.

Epiduralograms: Contrast radiographic study of epidural space after injecting air or positive contrast agent.

- It is safe and useful in diagnosis of space occupying lesions of the spinal cord.

Discography: Examination of the intervertebral disk space using X-rays after injection of contrast media into the disk.

- Inject the contrast agent in the nucleus pulposus of intervertebral disc.
- If the disc is ruptured 2-5 ml water soluble medium is injected.
- In normal disc only 0.5-1 ml could be injected.

Vertebral venography: Injection of radiocontrast medium into the saphenous vein while the caudal vena cava is compressed causes the vertebral veins to be outlined; used to demonstrate cord compression.

- Vertebral venography is used to locate compression of the spinal cord, swelling of the spinal cord and neoplasia of the cord or vertebrae.
- For cervical studies, injection of the contrast agent in the angularis oculi vein or nasal or facial veins are used with compression of the jugular vein.

Cerebral angiography: A radiographic procedure used to visualize the vascular system of the brain after injection of a radiopaque contrast medium.

Indications

- Aneurysm, thrombosis, ruptured arteries.
- Space occupying lesions that causes displacement of the vessels.
- Arteriovenous fistula and neoplasms.

Procedure

- Anesthetize the patient.
- Ligate the external carotid artery cranially and the common carotid artery caudal to the proposed catheterization.
- Catheterize the common carotid artery and artery is ligated around the catheter.
- Flush the artery with heparinized saline and infuse the meglumine iothalmate @ 1 ml/3 kg body weight + 1 ml for catheter.
- Take lateral and VD projections at 1 second interval.

Angiography: To visualize the vascular system.

Pneumoencephalography: Radiographic visualization of the cerebral ventricles and subarachnoid spaces after the injection of air or gas

- Injection of air/oxygen in the lumbar subarachnoid spaces visualized the cranial subarachnoid spaces and ventricles. This is called pneumoencephalography.

Ventriculography: Contrast radiographic study of ventricles of brain after the ventricular fluid is replaced by air or by an opaque medium.

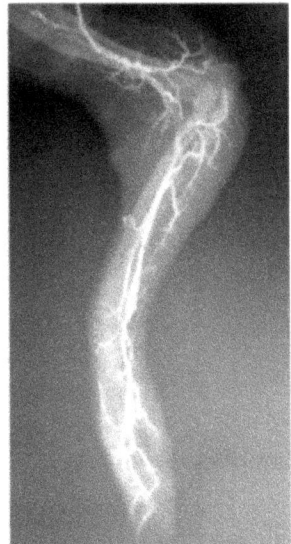

Fig. 14.5: Limb angiography in a rabbit

Indications

- Unilateral or bilateral ventricular dropsy.
- Space occupying lesions-tumors, hemorrhage, abscess.

Procedure

- Contrast agent (meglumine iothalmate) is injected tube inserted through trephine hole in the skull.
- Ventriculography requires general anesthesia whereas encephalography does not.
- Landmark for entry in to the lateral ventricle is approximately midway between the lateral canthus of the eye and the external occipital protuberance.

Cranial sinus venography: For visualization of structures of the cranial vault. It can be divided in to dorsal sagital sinus venography and cavernous sinus venography.

Indications

- Visualization of the orbit, ventral cranial sinuses and veins.

Procedure

- Cavernous sinus venography: By injecting contrast media in to the angularis iculi vein with compression of external jugular vein.

Fistulography: Contrast radiographic study of draining tracts like sinus and fistula.

Indications

- To identify the origin direction, path and extent of draining tract.
- To identify the radiolucent foreign body.

Procedure

- Take a catheter and fill the contrast media in it.
- Insert the catheter into the tract and pack the area around the catheter with gauze.
- Inject the water soluble triiodinated contrast material in to the draining tract and make a radiograph.

Complication

- Dissemination of infection.

Arthrography: Contrast radiographic study of the articular surfaces and joint capsule after injecting contrast media (positive or negative) into the joint space.

Indications

- Chronic joint distension.
- Evaluation of articular cartilage, subchondral bone and synovial membrane abnormalities.
- Joint capsule tearing.

Procedure

- Aseptic preparation of the site.
- Induce general anesthesia in small animals. In large animals inject the local anesthesia in the overlying soft tissues.
- Place the needle in the joint space. Aspiration of the synovial fluid confirms the correct placement of the needle.
- Withdraw the equal volume of the synovial fluid before injection of the contrast agent.
- Inject triiodinated water soluble contrast agent (4-5 ml in small animal and 5-20 ml in large animals) in the joint.
- Withdraw the needle and flex and extend the joint to mix the agent with synovial fluid.
- Make a radiograph immediately within 1 minute.

Tendonography: Contrast radiographic study of tendons.

Indications

- To outline the ligaments and tendons on the palmer/planter aspect of the distal 3rd metacarpal and metatarsal.
- Tendinitis, peritendinitis, desmitis and tendovaginitis of the tendon sheath.

Procedure

- Inject the contrast agent (gas) in the digital sheath between the superficial and deep digital flexor tendons and subcutaneously.
- Make a radiograph.

Reticulography: Contrast radiographic study of reticulum after administration of positive contrast agents.

Indications

- Reticular hernia
- Pericardiophrenic adhesions
- Phrenic abscess
- Pleuricy

Procedure

- Off fed the animal for 24 hour before reticulography.
- Administer about 1 kg barium sulfate suspension orally.
- Make a left lateral radiograph of caudal thorax and reticular area after 30 – 40 minutes.
- Left lateral radiograph taken in dorsal recumbency is more informative.

Fasciagraphy: Contrast radiographic study of tendons and associated structures.

Indications

- To evaluate adhesions
- Rupture of tendons and muscles
- Calcification

Procedure

- Sedate the animal and restrain in the lateral recumbency keeping the affected limb lowermost.
- Apply rubber tourniquet above and below the area of interest.
- Insert the 16 gauze hypodermic needle subcutaneously between the tendon and muscles. Attach a 3 way stopcock and inject the air by large syringe until the area is moderately distended.
- Withdraw the needle and make a radiograph.
- Remove the air and release the tourniquet.

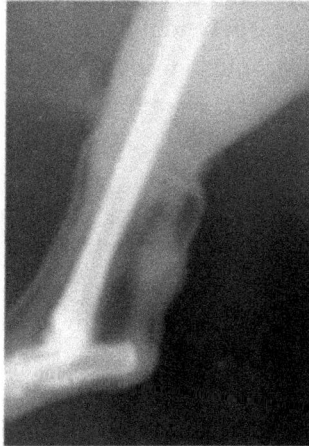

Fig. 14.6: Fasciagraphy of achillis tendon to evaluate healing and adhesions in a dog

Osteomedullography: (Intravenous phlebography): Contrast radiographic study of the intraosseous and extraosseous venous channels of the long bones.

Indications

- To diagnose delayed and nonunion.
- To observe the fracture healing.

Procedure

- Anesthetize the patient.
- Place the animal in lateral recumbency with the affected limb lower most.
- Aseptic preparation of medial aspect of metaphyseal region.
- After skin incision, a hole is drilled in to the cortex of the bone.
- Place the spinal needle in the hole and flush with 15-20 ml of normal saline.
- A rubber tourniquet is applied proximal to the drilled bone.
- 10-20 ml of water soluble iodine based contrast agent is injected rapidly through the needle.
- At the end of the injection make a lateral radiograph at 2, 4 and 6 minutes.
- Remove the tourniquet and take radiograph at 10, 20 and 30 minutes.
- Remove the needle and close the skin wound routinely.

Vaginography: Contrast radiographic study of vagina after administration of contrast agent into the vagina.

Indications

- To evaluate the vaginal masses
- Ectopic ureter and urethral tumor

Celiography: Is done for evaluating the abdominal cavity and diaphragm.

Indication:

- Diaphragmatic hernia

Procedure

- Make a survey radiograph.
- Aseptic preparation of the site (slight right to the midline and caudal to the umbilicus).
- Anesthetize the patient.
- Inject contrast medium (water soluble organic iodide) in to the abdominal cavity @ 2 ml/ kg body weight.

- Roll the animal.
- Make lateral and VD radiograph.

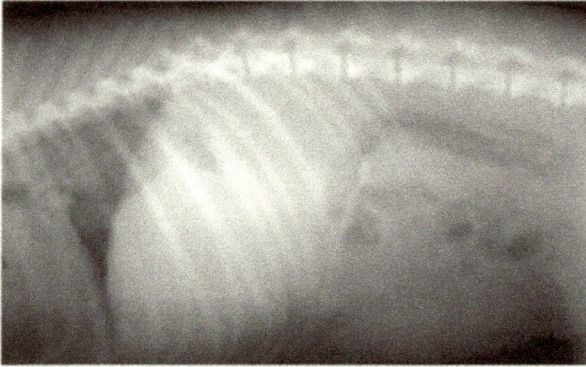

Fig. 14.7: Celiography in a dog

15

Principles of Viewing and Interpretation of X-ray Films, Classification of Radiographic Lesions

PRINCIPLES OF VIEWING

- Radiograph should be viewed in a darkened room,
- At least two viewing boxes are required.
- A bright light illuminator can be used for relatively over exposed areas.
- For detecting of normal radiographic anatomy or any abnormalities, the radiograph should always be placed on the illuminator as follows-
1. Lateral view (L) of axial skeleton: **(Fig. 15.1)**
   - Cranial (rostral) aspect of the animal to the radiologist's left.
2. Ventro-dorsal (VD) or dorso-ventral (DV)view: **(Fig. 15.2)**
   - Cranial (rostral) aspect of the animal pointing up.

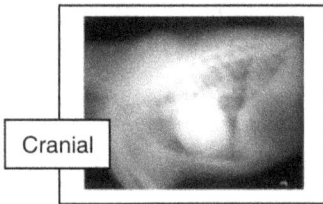

Fig. 15.1: Lateral View   **Fig. 15.2:** Ventro-dorsal (VD) view

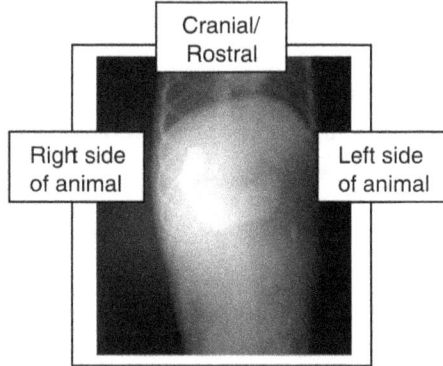

3. Latero-medial or medio-lateral view of appendicular skeleton: **(Fig. 15.3)**
   - Proximal extremity of the limb pointing up.
   - Cranial or dorsal aspect of the limb to the radiologist's left.
4. Caudo-cranial (palmo-dorsal and planto-dorsal) or Craniocaudal (dorso-palmer and dorso-planter): **(Fig. 15.4)**
   - Proximal end of the extremity at the top.
   - Lateral aspect of the limb to be placed on the radiologist's left.

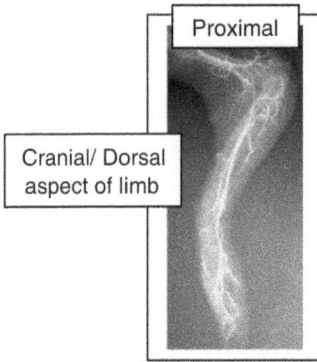

Fig. 15.3: Latero-medial or
medio-lateral view

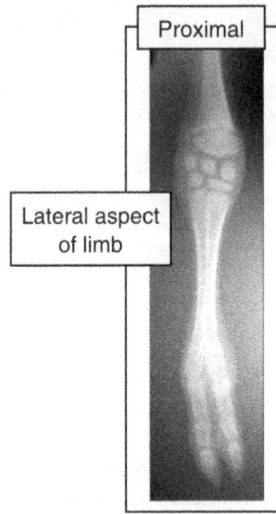

Fig. 15.4: Caudo-cranial
view

RADIOGRAPHIC INTERPRETATION

It is based on the visualization and analysis of opacities on a finished radiograph. For interpretation of a radiograph-

1. Take two radiograph of the part at right angle to each other.

2. The radiograph should be of good detail, correct density and proper scale of contrast.

3. The radiograph should be well dried.

4. The radiographic image is two dimensional images. The third dimension should be reconstructed in the brain from two radiographic projections made at right angle to each other.

5. Radiography of normal parts/organ should be used as reference.

6. Before final diagnosis, the lesions should be correlated with the history and clinical signs and laboratory data.

Radio-opacity

- Radio-opaque tissue/object- whiter image.
- Radiolucent tissue/ object- Black image.

The radio-opacity depends on-

1. Atomic number: Tissue or objects of high atomic number are more radio-opaque.

2. Physical opacity:

S.No.	Tissue/object	Effective atomic number	Specific gravity	Radio-opacity
1.	Gas (In trachea)	1-2	0.001	Black (Most radiolucent) **Fig. 15.5**
2.	Fat	6-7	0.9	Gray
3.	Soft tissue/ fluid	7-8	1.0	Gray
4.	Mineral (bone) Ca and P	14	1.8	White **Fig. 15.6**
5.	Metal or Lead	82	11.3	White **Fig. 15.7**

(*i*)   Gas opacity

- The gas is most radiolucent and provides negative contrast to visualize the various structures. *e.g.* the heart and vessels are outlined against the air filled lung in the thoracic cavity.

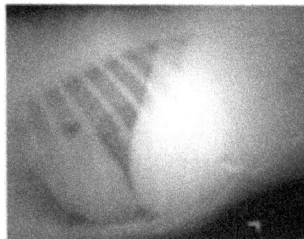

Fig. 15.8: Radiolucent gas around the heart and vessels

(*ii*)   Fat opacity

- Fat is more radiolucent than soft tissue or bone but is more radio-opaque than gas.

(*iii*) Soft tissue/fluid
- Have same radio-opacity e.g. heart, spleen, liver etc.
- Variation in thickness and degree of compactness of soft tissue creates a pattern of various densities on the radiograph.

(*iv*) Mineral/bone opacity
- Radio-opacity within same bone differs due to compact Vs spongy bone, Travecular bone Vs inter-travecular spaces, cortical bone Vs medullary canal.
- In disease condition, the bone may be more radio-opaque (sclerotic) or less radio-opaque (porotic) than normal bone.

(*v*) Metal opacity
- Most opaque shadow is formed on the radiograph.
- Contrast media: Barium sulfate, iodine containing contrast agents.
- Orthopedic implants and metallic foreign bodies.

3. Thickness
- Thicker tissue/object attenuates more X-rays and hence whiter image is formed.
- When tissues/objects are superimposed, the image will be more opaque.

SUMMATION SHADOWS

It results when parts of a patient or object in different planes (*i.e.* not in contact with each other) are superimposed.

1. Radiolucent summation shadow: When a block of Swiss cheese is radiographed, fewer X-rays are absorbed by the cheese. In the areas where the cavities overlap (Swiss cheese effect) more X-rays will reach the film.

2. Radio-opaque summation shadow: When a bunch of grapes is radiographed, more X-rays are absorbed by the areas where many grapes overlaps (bunch of grapes effect) *e.g.* visibility of miliary pulmonary metastasis.

SILHOUETTE EFFECT/ BORDER EFFACEMENT

When two structures of the same radio-opacity are in contact with each other, their individual margins at the point of contact cannot be distinguished (e.g. Liver and stomach are in close contact) and composite shadow is formed on a radiograph.

Border effacement results in diseases like pleural effusion, pooling of fluid around the heart (In DV radiograph, the heart margins are invisible).

EVALUATION OF THE RADIOGRAPH

- Determination of an abnormality in the radiograph.
- Find out the anatomic location of the abnormality.
- Classify the abnormality according to roentgen sign.
- Make a **Gamut** (list of differential diagnosis) e.g. If the kidney is enlarged, the possible gamut are- neoplasia, cyst, abscess, hydronephrosis and subcapsular urine/hemorrhage.

Roentgen signs: Radiologic abnormalities of tissues/organs/objects.

1. Abnormality in the size of an organ or structure.
    - (*i*) Increase in size: e.g. neoplasia
    - (*ii*) Decrease in size: e.g. Poor development of liver with porto-systemic shunt.
2. Variation in number of organs: e.g. absence of kidney
3. Variation in contour/shape: e.g. hypertrophy of heart in cardiomyopathy.
4. Change in position of an organ or structure: e.g. presence of abdominal organs in the thoracic cavity in diaphragmatic hernia.
5. Change in opacity of an organ or structure in certain conditions-
    - (*i*) Fluid filled tympanic bulla: Increased radio-opacity.
    - (*ii*) Calcification of soft tissues: Increased radio-opacity.
    - (*iii*) Radio-opaque foreign bodies: Increased radio-opacity.
    - (*iv*) Subcutaneous emphysema: More radio-lucent.
    - (*v*) Osteoporosis, osteomyelitis of bone: More radio-lucent.
6. Change in normal function of the organ: By using contrast agents.

Radiographic diagnosis: consist of two parts-

1. Location of the lesion in the body.
2. Classification of the radiographic lesion: Radiographic lesions can be –
    - (*i*) Congenital anomalies/developmental
    - (*ii*) Traumatic condition
    - (*iii*) Infectious condition
    - (*iv*) Malignancy/neoplasia
    - (*v*) Nutritional/metabolic
    - (*vi*) Hormonal imbalance
    - (*vii*) Degenerative lesion.

Radiation ionizes intracellular water and releases toxic products. These products can damage cell/DNA.

Radiation Sickness

When X-ray beam traverses through the tissues, the cells come to a state of high chemical reactivity which can initiate biological effects. The patho-physiological changes caused by cellular injury leads to radiation sickness.

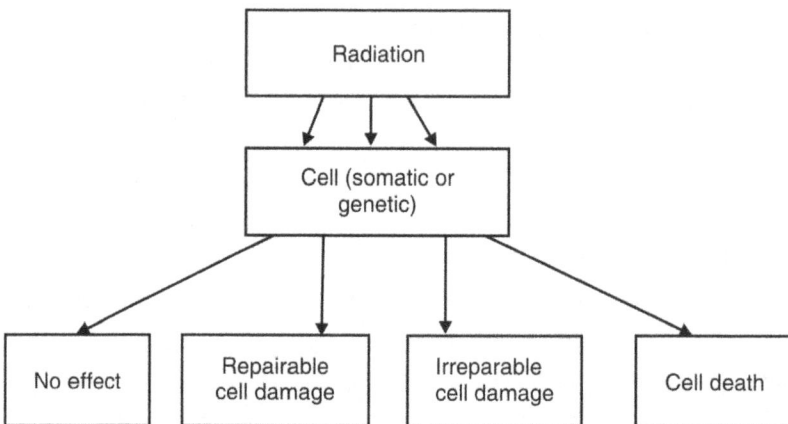

- Somatic cell damage affects the person in his life time. Leukemia and malignant tumors are common somatic effects.
- Genetic cell damage affects the subsequent generation.
- Pyrimidines (thymine, cytosine & uracil) are more radiosensitive than purines (adenine & guanine).
- Thymine is the most sensitive pyrimidine.
- Radiation sensitivity refers to the loss of reproductive capability of the proliferating/dividing cells.
- Rapidly dividing/growing cells are most sensitive to ionizing radiation.
- The fatty acids are attacked by free radicals at the double bond between carbon atoms.

- The carbohydrates are attacked by depolymerization.
- The cells which do not proliferate are relatively radioresistant e.g. nerves and muscles.
- Stem cells of the hematopoietic system, cells of the gut, skin and testis are highly radiosensitive.
- The cells can be classified on the basis of radiosensitivity as follows-
    1. Vegetative intermitotic cells: These cells are rapidly dividing cells and are highly radiosensitive e.g. megaloblasts and germinal layer of the skin.
    2. Differentiating intermitotic cells: These cells are daughter cells of vegetative cells and are less radiosensitive e.g. promyelocytes.
    3. Post mitotic cells: The cells do not divide or divide at low rate and are relatively radioresistant e.g. renal epithelial cells and hepatic cells.
- Lymphocytes are most sensitive hematopoietic cells.
- Thyroid gland, intestinal epithelium and eye lenses are sensitive to radiation.
- LD50/30: It is the dose that will kill 50% of the population within a period of 30 days. LD50/30 is affected by the source of radiation and dose rate and it is highest for rabbits (7.5Gy).
- Chronic low level skin exposure may lead to squamous cell carcinoma and radio dermatitis (reddened dry skin).
- Radiation exposure of more than 25 Rads (0.25 gray) to developing fetus leads to significant damage as follows-

S.No.	Time of Radiation exposure	Effect on fetus
1.	Pre-implantation period (0-9 days)-	Death of the embryo
2.	Organogenesis period (10 days to 6 weeks)	Congenital malformations
3.	6 weeks to end of the term	Least sensitive time. Slow growth and mental retardation may occur

Acute radiation sickness (Early effects of radiation)

The symptoms appear within days or weeks after exposure to radiation. Exposure to intense radiation, the radiosensitive cells of the entire body system is severely damaged simultaneously. This may lead to immediate severe lymphopenia and damage to the stem cell pool. The combined effect is called as radiation sickness. The death caused by radiation damage to hematopoietic system is the result of severe neutropenia which permits the development of fulminating infection in the body.

UNITS OF RADIATION

1. **Electron volt (eV) and Kilo electron volt (keV):** Unit for measurement of energy of an X-rays.

2. **Roentgen (R):** It is the quantity of an X-rays or gamma radiation which produces one electrostatic unit (2.08×10^9 ion pairs/cm^3) in 1 cc of dry air after its ionization at 0^0C and 760 mm Hg.
   - It is the unit for exposure to X-rays or gamma rays.
   - It measures the quantity of ionization and does not describe the dose to tissues.

3. **Roentgen equivalent man (rem):** It is the amount used to express the dose equivalent that results from exposure to ionizing radiation. (1 mrem = 0.001 rem).
   - 1 Rem= Rad x quality factor
   - Quality factor relates to the biological effectiveness of the given radiation.
   - Quality factor for X-rays and gamma rays is 1 and for alpha particle is 20.

4. **Sievert (Sv):** In SI system the unit of dose equivalency is sievert.
   - Sv = absorbed dose(in gray) x weighting factor
   - 1 Sv = 100 rem

5. **Gray (Gy):** It is defined as the amount of radiation such that the absorbed energy is 1 joule/kg of tissue.
   - It is the SI unit for absorbed dose
   - It is used to define rad
   - 1 Gy = 100 rads

6. **Radiation absorbed dose (Rad):** It is the unit of absorbed dose following exposure to any ionizing radiation.
   - One Rad is equal to the radiation necessary to deposit energy of 100 ergs in 1 gram of irradiated material.
   - Rad and Roentgen are approximately same when the soft tissues are exposed to X-rays or Gamma rays.

ALARA (As Low As Reasonably Achievable) principle: It is the official method of choice for limiting exposure to radiation.

ALARA MPD: The MPD for non occupational person is as low as 0.1 rem/year or 1mSv/year.
   - Fetus should not receive >0.5 mSv or 0.05 rem/month during entire gestational period.
   - Pregnant staff member should wear an additional badge at waist level underneath the lead gown to monitor the fetal dose. This badge should not exceed 0.05 rem /month.

> **BIER:** Biological effects of ionizing radiation
> **USNRC:** United States nuclear regulatory commission
> **NCRP:** National council on radiation protection

EXPOSURE SOURCES

1. **Primary beam**
    - Protective devices having 0.5 mm of lead can decrease the primary beam exposure by 25% only.
    - Protective devices are used to protect the body from scatter radiation.
2. **Scatter radiation:** Scatter radiation can be defined as the X-rays degraded by collision within tissue molecule,
    - It may come from the patient, floor and sides of the table.
3. **X-ray tube head:** Occasional leakage of the X-rays from the tube housing.
4. **Fluoroscopy:** During fluoroscopy approximately 5000 m rem/ minute is emitted in the actual beam.

Minimizing exposure

1. **Lead shielding**
    - All the staff members and patient's owner should wear the protective devices (Lead apron, gloves, goggle etc.) while exposure is being made.
    - Lead apparel should be hanged or draped over the rounded surface without folds or wrinkles to prevent damage to the lead lining.
    - All protective devices should be screened (5mAs and 80 kVp) every 6 months
2. **Increased distance**
    - Exposure to staff can be minimized by increasing distance from the primary beam (6 feet).
    - As per the law of Inverse Square, intensity of the x-ray beam decreases to $1/4^{th}$ if the distance between the x-ray source and operator is doubled.
    - During exposure, the restrainers should lean back and look away from the beam to protect their eyes.
3. **Reduced time**
    - Use of fastest film-screen combinations results in reduced exposure time to patient and personnel.
4. The beam should be collimated down to the area of interest to reduce scatter radiation exposure to staff.

5. Cones and diaphragm attached to the tube window also reduces the scatter radiation.

6. A 2mm aluminum filter is used at the tube window to filter the soft rays.

RADIATION MONITORING DEVICES

- Radiation monitoring devices are used to record the amount of radiation received by the person during radiographic procedures.
- These devices are worn during radiographic procedures and then are sent to an approved laboratory to be processed.
- Four main type of monitoring devices are used-
   1. Film badges
   2. Thermo luminescent dosimeters (TLDs)
   3. Optically stimulated luminescence (OSL) badge
   4. Ion chambers

1. Film badges

- Most commonly used monitoring device.
- Film badge contains a small piece of unexposed film enclosed in a light proof envelope.
- The badge is placed at the collar or wrist.
- The badge is worn for a specified period and then sends to the laboratory.
- The film is developed and the value in mrem is correlated by the degree of blackness on the film.
- The amount of radiation received is directly proportional to the degree of blackness on the film.

2. Thermo luminescent dosimeters (TLDs)

- It contains lithium fluoride or calcium fluoride crystals. These crystals store the energy produced by radiation exposure.
- The crystal undergoes heat processing and a light is emitted.
- The amount of light emitted is directly proportional to the amount of radiation that dosimeter received.
- Thermo luminescent crystals records greater amount of energy than the film badge.

3. Optically stimulated luminescence (OSL) badge

- A thin layer of aluminum oxide is stimulated with different frequencies of Laser light. Illumination of aluminum oxide is proportional to the amount of radiation received.
- OSL badge are extremely sensitive to the radiation.

4. Ion chambers

- These devices are the size of a pen and can be clipped on to a pocket.
- An electrometer is used to charge the chamber before being worn. After each exposure, the ion chamber discharges. The amount of discharge of the ions is proportional to the amount of radiation received.

GENERAL SAFETY MEASURES: During radiography

- Always wear lead apparels during radiography in the exposure room.
- Always wear a radiation monitoring device on the collar outside the apron.
- Don't expose the body parts to primary beam
- Always look away from the X-ray beam during the exposure.
- Use restraining devices like anesthetic drugs, tape, sand bags, cassette holders etc.
- Pregnant women and persons under the age of 18 years should not be allowed in the exposure room.
- Minimum no. of persons should remain in the exposure room.

17 Principles of Ultrasonography and its Application in Veterinary Practice

Ian Donald (1957) developed first diagnostic ultrasound machine. Ultrasound is defined as sound wave of frequencies greater than which is audible to the human ear *i.e.* greater than 20,000Hz.

PROPERTIES

- Ultrasound is a high frequency sound wave.
- For diagnostic application 1-10 MHz frequencies are used.
- Ultrasound cannot propagate in a vacuum. In gas transmission of ultrasound is poor.
- Reflection of ultrasound occurs between substances of different acoustic impedance.
- Transducer and the patient surface must be bridge by a suitable coupling agent.

PRINCIPAL

The ultrasound beam (pulse) of specific frequency is generated by the piezoelectric crystal that oscillates within the transducer (probe). The pulse interacts with tissue in the body and reflected back in the form of echo which is received by the probe (pulse echo principal). The received impulse is converted to an electrical signal and processed by the computer to form a final image to display (echo quantification).

DOPPLER SHIFT

Difference between transmitted and received sound frequencies is known as Doppler shift. The greater the Doppler shift, the greater the flow velocity.

Interaction of ultrasound beam with tissues: It includes reflection, absorption and refraction.

1. **Reflection:** Used for diagnostic ultrasonography. Reflection means return of a portion of sound beam to the probe to form a final image.
   - Lungs: Poorest sound transmission due to very low density of air.
   - Bone: Rapid sound transmission due to high density.

- Other tissues: Intermediate sound transmission due to intermediate density.

- The strongest echoes are produced between tissues having the greatest difference in acoustic impedance. The bone and bowel air does not allow the scanning of tissues deeper to these tissues and nothing is seen beyond this interface.

- The echo generated by the reflection of ultrasound beam at the tissue interfaces (junction of two tissues with different acoustic impedance) is of the following types-

    (*i*) A strong echo appears white - Hyperechoic.

    (*ii*) Intermediate strength echo - Echoic (light or medium shades of gray).

    (*iii*) Weak echo – Hypoechoic (dark gray).

    (*iv*) Absence of an echo- Anechoic (black).

2. **Absorption:** Absorbed ultrasound beam is converted to heat which is used as therapeutic ultrasound waves.

3. **Refraction or scattering:** It causes loss of image clarity and is related to the ultrasound artifacts.

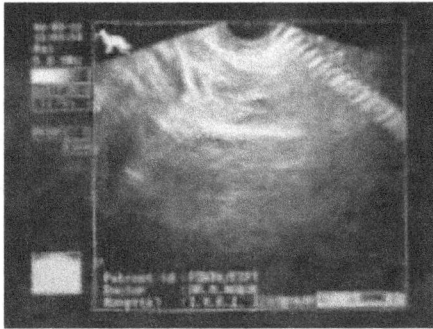

Fig. 17.1: Ultrasonographic image of ventral hernia

MODE OF DISPLAY

1. **A-mode (amplitude mode)**

   - Earliest and simplest form of ultrasound.

   - The returning echoes are displayed as a series of peaks on a graph. The distance from the transducer is depicted on horizontal line and amplitude on vertical line.

2. **B-mode (brightness mode)**

   - B-mode uses bright pixels/dots on the screen which corresponds to the depth at which the echo was formed.

   - Degree of brightness is directly proportional to the intensity of the retuning echo.

- This mode is currently in use for diagnostic ultrasound. It is if 2 times-
    - (*i*) Static B-mode: The probe/ transducer/sound beam is moved in the scanning plane by hand.
    - (*ii*) Real time B mode: The sound beam is automatically and rapidly moved in the scanning plane. Two dimensional images are continually updated.
- Low frequency transducers (3.5 MHz and 5 MHz) are used for scanning deep tissues but resolution is less.
- High frequency transducer (7.5 MHz) is used for scanning superficial structures. Such transducer produces better resolution of the image.
- This is because attenuation of sound beam is proportional to its frequency.
- High frequency gives better resolution than a low one.

ULTRASOUND TRANSDUCERS

- The transducer emits a series of sound pulses and received the returning echoes.
- The frequency of the transducer depends on the contraction and expansion of the crystal in one second.
- The configuration of the transducer may be mechanical or electronic. The scan plane can be either sector (pie shaped image) or linear array (rectangular shaped image).

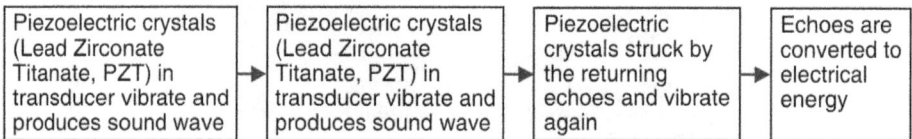

| Piezoelectric crystals (Lead Zirconate Titanate, PZT) in transducer vibrate and produces sound wave | → | Piezoelectric crystals (Lead Zirconate Titanate, PZT) in transducer vibrate and produces sound wave | → | Piezoelectric crystals struck by the returning echoes and vibrate again | → | Echoes are converted to electrical energy |

Types of transducer

1. **Linear:** Made up of multiple thin rectangular clips lined up side by side each producing sound wave. Rectangular shape image is created with good visualization of superficial structures.
2. **Convex:** The crystals are placed in a curvilinear fashion. Imaging of a greater area can be effected with the same contact area.
3. **Micro-convex/sector:** This type of transducer contains a single crystal which oscillates or rotates to produce a fan shaped beam and produces a wedge shaped image. It gives more accessibility to the thoracic and abdominal organs through a small contact area. The superficial structures however are not well visualized.

Image interpretation: Images are displayed as white against a black background. The image produced is a mixture of different echoes depending on the area scanned.

1. **Hyperechoic or echogenic:** Bright echoes which appear white on conventional scan e.g. images produced by highly reflective interfaces such as bone and air.

2. **Hypoechoic:** Gray images produced by interfaces of moderate reflection such as soft tissues.

3. **Anechoic or echolucent:** Black image produced by complete transmission of sound wave such as through fluids.

Contrast agents used

1. Microspheres of human albumin
2. Microparticles of galactose
3. Perflenapent
4. Microspheres of phospholipids
5. Sulfur hexafluoride

Scanning controls

1. **Time compensated gain**
   - Intensity of sound beam is decreased by approximately 1 decibel / cm / MHz. Thus unequal echoes will be bouncing back by equal acoustic interface located at different distance from the transducer.
   - TCG controls allow equal acoustic interfaces to return equal echoes regardless of the depth.
   - If the far field echoes are weak- TCG turned up.
   - If the near field echoes are weak-TCG turns down.

2. **Gain:** It determines the overall density of the picture.

Types of ultrasonography

1. **Intra-operative ultrasonography:** used to assist surgical procedures like
   - To determine margins of lesion.
   - To determine specific anatomical structure e.g. bile duct.
   - To determine flow patterns within vessels.

2. **Laparoscopy ultrasonography**
   - Refers to the insertion of a fixed or flexible ultrasound transducer through a laparoscopic port.
   - Used for direct imaging of intestinal structures.

3. **Endovascular ultrasonography**
   - A small ultrasound transducer is incorporated into the catheter and introduced intravenously to specific sites.
   - Cardiac evaluation with use of endovascular ultrasound catheter can provide exceptional image of the heart.
   - Used to monitor the effects of angioplasty and quantitating the degree of arterial stenosis.

4. **Endoscopic ultrasonography (endosonography)**
   - It refers to use of small high frequency transducer incorporated in to the tip of an endoscope, providing excellent near-field resolution.
   - The ultrasonographic endoscope is introduced in to the body cavity per-orally, transrectally or transvaginally.

Artifacts

1. **Acoustic shadowing**
   - Echo free zone created distal to the imaged organ when sound wave hit a highly reflective tissue that prevents sound from being transmitted to greater depth.
   - The most common acoustic shadows of clinical significance are produced by cystic, renal and biliary calculi.
   - Gas causes near total reflection of the beam.
   - Bone, gas and mineral deposits reflect the sound waves completely and bright image is produced with no visible structure beneath it.

2. **Acoustic enhancement or distant enhancement**
   - Fluid gives an anechoic image as the sound beam passes uninterrupted. A normal bright area of increased sound intensity is produced beyond a fluid filled space. This phenomenon is called acoustic enhancement e.g. during normal scanning of urinary bladder.

3. **Reverberation:** Process of echo bouncing back and forth between the two interfaces is known as reverberation. The presence of air between the probe and the patient leads to reverberation artifact.

4. **Mirror image:** Highly reflecting interfaces act like a mirror and shows mirror image. This defect is common during scanning of tendons.

5. **Comet tail:** Caused by highly reflective interfaces specially the air fluid interface e.g.
   - (*i*) In partially consolidated lung at the Interface between diaphragm and lung.
   - (*ii*) Interface between the bowel wall and bowel gas.

6. **Side lobe artifact:** It refers to lateral displacement of structures produced by minor beams of sounds travelling out in directions different from the

primary ultrasound beam. Curved surfaces like diaphragm, gall bladder, highly reflective surface etc. are common example of this artifact.

7. **Slice thickness artifact:** This artifact mimics the presence of sediment in the gall bladder or urinary bladder (pseudosludge- has a curved interface). True sediment usually has a flat interface and settles at the dependent portion of the organ while changing the position of the animal.

DOPPLER ULTRASOUND

It can be used to record blood flow velocities within the blood vessels

1. **Pulsed wave Doppler:** Blood flow velocities can be recorded from a specific location. It is superior to continuous wave Doppler because it gives information about the depth and velocity.

2. **Continuous wave Doppler:** Blood flow velocities can be recorded along a specific line. It is the simplest method.
    *   Two transducers are present in the same scanner. One to emit ultrasound and another to detect the returning echoes.
    *   Doppler shift: The difference in frequency between the emitted and incoming signals is known as Doppler shift. It is recorded to obtain velocity information.

3. **Colour flow Doppler:** The direction of flow-
    (i)   Positive (above the base line) or red: When blood flow is towards the transducer (arteries).
    (ii) Negative (below the base line) or blue: When the flow is in opposite direction (vein).

4. **Duplex:** In duplex Doppler pulsed wave Doppler unit is simultaneously used with real time sector imaging. This combination locates correctly the area of interest and then freezes it on the screen. Flow information is obtained from only a small part of the total image.

    *   Long axis view: Echocardiographic image showing the heart from base to apex in a longitudinal or Sagital plane.
    *   Short axis view: Echocardiographic image showing the heart in a transverse plane.

DIGITAL RADIOGRAPHY

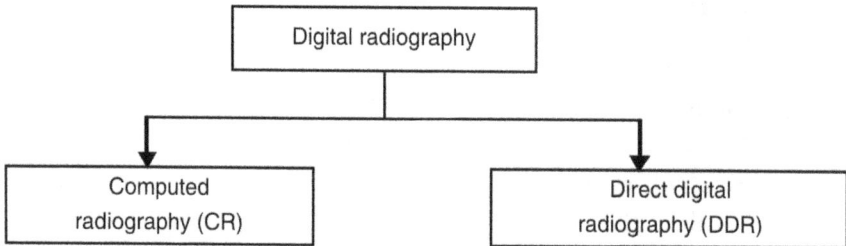

```
                ┌─────────────────────────┐
                │    Digital radiography   │
                └─────────────────────────┘
                            │
            ┌───────────────┴───────────────┐
            ▼                               ▼
┌─────────────────────┐         ┌─────────────────────┐
│     Computed         │         │    Direct digital    │
│  radiography (CR)    │         │  radiography (DDR)   │
└─────────────────────┘         └─────────────────────┘
```

1. Computed radiography

- An imaging plate which is stored in a cassette is used to record the latent image.
- An imaging plate is coated with photostimulable phosphors.
- The imaging plate is stroked by X-rays and electrons are energized to a higher energy state and stored in electron traps forming a latent image.
- The exposed cassette is placed in the plate reader. The plate reader scans the plate with red laser (helium-neon laser).
- In plate reader the electrons that were trapped in a higher energy state are released in to a lower energy state. In this event, phosphorescence (visible light) is produced which is detected by photomultiplier tubes in the plate reader.
- The light energy is amplified and converted to an electrical signal that is proportional to the light intensity released from the plate.
- This electrical signal (analog) is converted to digital data (numbers) by analog to digital conversion (ADC).
- The digital data is transferred to computer and image is produced or printed on film.
  - The image represents the collected data for each specific pixel that has a shade of gray corresponding to the degree of attenuation of the imaged part.
  - As the pixel size decreases, image resolution is increases.

2. Direct digital radiography

- The imaging plate is designed with an array of detector elements and built directly into the X-ray table (No need of cassette).
- The detectors are of 2 types-
  - (*i*) Direct: X-ray energy is directly converted to electrical signal.
  - (*ii*) Indirect: First the X-ray energy is converted to light and then converted to electrical signal.
- The digital image is produced within 4-10 seconds for immediate viewing.

Once image is produced, the image can be modified by the process *post-processing*. The post-processing changes are important to allow for variation in contrast.

1. **Windowing:** windowing is achieved through variation in the window width and window level (window center). The window width determines the range of exposure that can be assigned to a shade of gray. Pixels above the range will all be white, whereas pixels with a value below the range will be black. The window level represents the center point of the window width.
   - For assessing the bone high contrast window is needed.
   - For assessing the soft tissue low contrast window is needed.

2. **Smoothing:** It is a spatial filtering operation by which the adjacent pixels are blended to create a more uniform image without a large change in information between adjacent pixels. Smoothing can decrease the mottled appearance of a digital image. If too much smoothing occurs, the image can be blurred and detail is lost.

Increase in bit depth increases the number of gray shades/ pixel

Bit depth (n)	Shades of gray (2^n)
1	2
2	4
3	8
8	256
16	65536

Image quality

- Image quality depends on proper patient positioning and beam angle.
- Spatial resolution depends on matrix or pixel size. For digital radiography, a 2048 x 2048 (2K) matrix or higher is preferred because of the need for higher spatial resolution.
- Spatial resolution of CR images should be at least 2.5 line pairs/ mm and 10 bits/ pixels or greater.

Advantages of CR and DDR systems

- Linear response to X-ray intensity over wide latitude (dynamic range). Image contrast and latitude (range of exposure) are major advantages of digital radiography. Wide range of exposure factors that can be used without compromising the diagnostic value of the images.
- The latitude of digital radiography is much greater than that of film screen system
- Digital image can be adjusted to display high contrast (few shades of gray) or wide latitude (many shades of gray). Contrast and blackness can be adjusted after the image has been acquired by the process known as windowing.
- In digital system, the mAs setting is not related to blackness as with film screen system.
- Post processing is impossible in the film screen system.
- No need to take more radiographs due to technical error (under exposure or over exposure). Digital radiography compensates for improper exposure.
- Total time is decreased significantly.
- Computed image enhancement provides both edge and contrast enhancement.
- Digital storage and images can be viewed at distant locations.

Disadvantages

- Severe under exposure will result in a mottled or coarsely stippled image.
- Spatial resolution is better with film screen radiography. CR system has a spatial resolution of approximately 2.5 line pairs/mm whereas film screen system ranges from 6-10 line pairs/mm.

Artifacts in digital radiography

- If underexposure occurs, the resultant digital image appears grainy (pixilated). In an over exposed image thin, soft tissue structures are not visible.
- Improper use of edge enhancement may result in the lung fields appearing to have an interstitial pattern.
- The plate becomes worn over time. They can crack on the edges, resulting in white linear artifacts on the image.
- If the imaging plate is incompletely erased before the next exposure, the portions of the previous image or scatter radiation may be present.
- If the light guide on the plate reader needs to be cleaned, a white line artifact will appear on the image.

- Lack of primary beam collimation and failure to center the area of interest on the cassette may result in artifacts.
- Any debris on an imaging plate will result in a sharp white artifact.
- Backscatter (images of object behind the cassette) can also be seen with CR plates.
- Low line grids will result in moiré pattern, if the grid lines are parallel to the reader's scan lines.

PACS (Picture, Archiving and Communication Systems)

RIS (Radiology Information System)

DICOM (Digital Imaging and Communications in Medicine)

COMPUTED TOMOGRAPHY

Synonyms: X-ray computed tomography/Computed axial tomography (CAT or CT scan)

Tomography: Imaging by section

Tomogram: Image produced by tomography

Pixels/picture elements: tiny squares make up the image matrix. The two dimensional image that is produced is composed of many squares called pixels.

Voxels/volume elements: Three dimensional box represented on an image matrix by the two dimensional pixels.

Principle: Very fine X-ray beams are sent through the body to detectors. The detectors then send the signals to a computer which processes the image.

- Invented by Sir Godfrey Hounsfield (1979)-got novel prize for medicine
- It is a non-invasive technique.
- Body structures can be visualized in dorsal, sagital, oblique and transverse planes without superimposition artifacts from fat, bones or any organs that may mask the detail on a survey radiograph.
- CT provides better differentiation of soft tissues than conventional X-rays, since the gray scale can be controlled by the operator.

CT unit: Consist of

1. A movable bed or cradle: The anesthetized patient lies on it. The cradle moves in gantry during scanning.
2. Gantry: It is a ring that contains X-ray tube that is positioned on the opposite side of the detectors. It can be moved 360^0 around the patient.
3. The earliest sensors were scintillation detectors with photomultiplier tubes excited by cesium iodide crystals. In new machines cesium iodide crystals are replaced by ion chambers containing high pressure xenon gas.

Procedure

- Patient is placed (DV or VD or Lateral) in the gantry.
- Switch on the machine. Numerous x-ray beams and a set of electronic x-ray detectors rotate around the patient. At the same time cradle moves through the scanner.
- An X-ray detector absorbs the photons emerging from the patient and converts these to electronic signals.
- Electronic signals are assigned a number which represents their intensity.
- Computer in separate room reconstructs the information into a picture displayed on a television screen.
- CT scanning lasts between 5 to 30 minutes.
- A set of images or slices are produced at each interval through the gantry.
- The computer reconstructs the internal structure from several projections of the organ.

CT number/ Hounsfield number: The density of each voxel is compared with density of water and then assigned to a gray scale shed. It represents the attenuation of the X-ray beam in tissue within a voxel.

- CT number for

    1. Metal: +3000
    2. Bone: +1000
    3. Air: -1000
    4. Water: 0

Windowing: CT produces a volume of data which can be manipulated through windowing in order to demonstrate various structures based on their ability to block the X-ray beam.

Indications

- To analyze the internal structures of various body parts like head injury (traumatic, blood clots, skull fracture), tumors and infections.
- In the spine structure of the vertebrae, intervertebral disc and spinal cord can be accurately defined.
- To measure the density of the bone in evaluating osteoporosis.
- Diagnosing cancers and plan of surgery: To confirm the presence of tumor measure its size, precise location and extent of tumors involvement.
- To take guide biopsies.

Advantage over traditional radiography

- It eliminates the superimposition of the images of structures.
- Because of the inherent high contrast resolution of CT, differences between tissues that differ in physical density by less than 1% can be distinguished.
- CT angiography avoids the invasive insertion of an arterial catheter and guidewire.

Disadvantages: Same as conventional X-rays.

Adverse reaction to contrast agents

- Nausea and discomfort.
- Life threatening allergic/anaphylactic reactions If reaction occurs medications like corticosteroids, antihistaminics and epinephrine should be used to reverse the reaction.
- Kidney damage: So iodinated contrast agents should be avoided.

Difference between CT and MRI

S.No.	CT scan	MRI
1.	It uses X-rays, (ionizing radiation).	Uses non-ionizing radiofrequency (RF) signals to acquire the image.
2.	Limited to axial plane in past. But with the use of multidetector CT scanners, image can be reconstructed in any plane.	MRI can generate cross-sectional image in any plane (including oblique plane).
3.	Contrast agents used: iodine and barium.	Agents with paramagnetic properties like Gadolinium.
4.	Can be performed in presence of a medical device.	Cannot be performed in presence of a medical device.
5.	Contrast is not as good as MRI.	MRI provides much greater contrast between soft tissues of the body than CT.
6.	CT provides good spatial resolution (The ability to distinguish two structures arbitrarily small distance from each other as separate).	MRI provides better contrast resolution (The ability to distinguish the difference between two arbitrarily similar but not identical tissues).
7.	For tumor detection and identification CT is not so good.	MRI is superior.

Artifacts

S.No.	Artifact	Definition	Cause	Remedy
1.	Aliasing artifacts or streaks	Dark lines which radiate away from sharp corners.	(i) Metallic object which can completely extinguish the x-rays. (ii) Insufficient tube current is selected.	(i) Remove the metallic object. (ii) Use proper tube current.
2.	Motion artifact	Blurring or streaking.	Movement of the object being imaged.	Restrict the movement of an object during scanning
3.	Partial volume effect	Blurring over sharp edges i.e. one part of the detector can not differentiate between different tissues	Scanner is unable to differentiate between a small amount of high density material (bone) and a large amount of lower density (cartilage).	Scan the thinner slices.
4.	Ring artifact	It is the most common mechanical artifact. The image of one or many rings appears within an image.	It is due to fault in detector.	Replace the detector.
5.	Noise artifact	Graining on the image.	(i) Cased by low signal to noise ratio and occurs when a thin slice thickness is used. (ii) It also occurs when the power supplied to the x-ray tube is insufficient to penetrate the object.	(i) Set the proper slice thickness. (ii) Use proper power supply.
6.	Windmill	Streaking.	When the detectors intersect the reconstruction plane	Use filters or reduce the pitch.
7.	Beam hardening	Cupped appearance.	When more attenuation in the center of the object than around the edge.	

MAGNETIC RESONANCE IMAGING (MRI)

Synonym: Magnetic resonance tomography (MRT)

It is the best non-invasive technique to visualize the soft tissue changes. The image is produced by applying nuclear magnetic resonance phenomenon. There is no known side effect of MRI scan on the body. MRI uses a strong magnetic field and radiofrequency (RF) wave to provide extraordinary details on the picture of internal organs and tissues.

History

- NMRI was converted to MRI to eliminate the word nuclear because the public had a strong aversion to the term "nuclear".
- Damadian (1974) patented the design and use of MRI for detecting cancer.

- Paul Laterbur and Sir peter Mansfield (2003) got Nobel prize in medicine for their discoveries concerning MRI.
- Paul Laterbur named this technique as Zeugmatography but this term was not adopted.

Units:

- Gauss (G) and Tesla (T). One Tesla = 10,000G
- Magnetic field strength of MRI machine ranges between 0.3 to 3 Tesla.
- Earth's magnetic field is 0.5 G.

Principle: MRI depends on the spinning motion of specific nuclei present in the tissue. The specific nuclei are known as MR active nuclei.

Components of MRI Scanner

1. For static magnetic field superconducting electromagnet is used.
2. An RF transmitter and receiver.
3. Three orthogonal, controllable magnetic gradients.

An atom is made of electrons (orbiting around the nucleus), proton and neutrons (Nucleus). MRI depends on the magnetic properties of excited hydrogen nuclei (protons) which moves randomly in water and lipids. When an external magnetic field is applied these protons line up parallel to the applied field when radiowaves (RF pulse) are applied to these protons. They absorb energy from the radio waves and flip to 90^0 from the applied magnetic field. Thereafter radiowaves are stopped. Protons gradually return from excited position to their parallel position and energy is released in the form of RF pulse which is grabbed by sensors in the MRI machine and is measured by the computer. Based on this information, computer forms well detailed image of the exposed organ **(Fourier transform)**.

Angular momentum (AM) is the rotational motion of the particle and AM of proton and neutron is opposite to each other. Some atoms have unpaired proton and neutrons. These atoms have net AM (spin). MR measures this spin. Hydrogen atoms are the most abundant unpaired atoms present in the body in the form of water (free form), proteins, fat (bound form), carbohydrates and nucleic acid. Hydrogen is the solitary proton gives the large magnetic moment and used as MR active nuclei. Contrast of fat and water appears differently in MR images.

- A tissue has high signal (white on the image): If it has a *large* transverse component of magnetization and amplitude of the signal received by the coil is *large*.
- Low signal (dark on the image) :): If it has a *small* transverse component of magnetization and amplitude of the signal received by the coil is *small*.
- Some tissues have intermediate signals (shades of gray between black and white).
- T_1 weighted images are characterized by bright fat and dark water.

- T_2 weighted images are characterized by bright water and fluid containing tissues and dark fat.
- Areas of higher proton density are bright and low proton densities are dark.
- Protons in water and other fluids (CSF and Synovial fluids) have long T_1 and T_2 relaxation times and proton in fat have short T_1 and T_2 relaxation times.
- Echo time (T_E) and Repetition time (T_R) are basic parameter of image acquisition.

T_1 relaxation: energy is released during return of excited nuclei from the high energy state to low energy. The realignment of nuclear spins with the magnetic field is termed as longitudinal relaxation and the time required for a certain percentage of the tissue's nuclei to realign is known as time 1 or T_1 (about 1 second).

T_2 relaxation: (spin-spin relaxation). When spins in high and low energy state exchange energy without loss of energy to the surrounding lattice. This results in loss of transverse magnetization. Transverse relaxation time is termed as time-2 or T_2 (100msec for tissues). In pure water, the T_1 and T_2 time are approximately same i.e. 2-3 seconds. In tissues, the T_2 time is shorter than T_1 time.

T_1 weighted MRI: (spin-lattice relaxation time). It is a basic standard scan used for differentiating fat (brighter) from water (darker). T_1 weighted scan uses a gradient echo (GRE) sequence with short T_E and short T_R. Due to short TR this scan can be run very fast. This scan is

T_2 weighted MRI: (spin-spin relaxation time). It is also a basic scan for differentiating fat (darker) from water (brighter). T_2 Image well suited for imaging edema. T_2 weighted scan uses a spin echo (SE) sequence with long T_E and T_R.

Larmor frequency: The frequency with which the dipole moments precess is called the larmor frequency.

Indications

- To distinguish pathological tissue from normal tissue
- To distinguish between arbitrarily similar but not identical tissues.
- FLAIR (fluid light attenuation inversion recovery) sequence useful for determining difference between free water (dark) and edematous tissue fluid (bright).

Contrast enhancement: MR contrast agents shorten the T_1 and T_2 relaxation times of different tissues. Contrast agents are used for diagnosis of tumors, infection, infarction, inflammation and post-traumatic lesion in the body.

(*i*) **Paramagnetic contrast agents:** used for brain perfusion, tumors etc. Gadolinium and Gadoxetate are positive contrast agent. Tissues appear extremely bright on T_1 weighted images.

(*ii*) **Super Paramagnetic contrast agents:** Iron oxide nano-particles. Used for liver and GI tract imaging. Tissues appear very dark on T_2 weighted images.

(*iii*)   **Diamagnetic agents:** Barium sulfate. Used for GI tract imaging.

Applications

1. **Neurology**
    - To visualize the brain affections like cerebral infarction, tumors etc.
    - To visualize the spinal cord and surrounding tissues like intervertebral disc diseases, tumors, spinal cord, spinal nerves, emboli and other pathological conditions of the spinal cord.

2. **Musculoskeketal system**
    - Tenosynovitis, rupture of the ligaments, menisci rupture and tendon injuries, arthritis and lameness in horses.

3. **Abdominal and thoracic studies**
    - To visualize the organs of the abdominal and thoracic cavity.
    - To diagnose metastasis from neoplastic process.

4. **Oncology**
    - To visualize the neoplasms.

5. **Cardiovascular:** e.g. MR angiography.

Advantages

- No use of ionizing radiation. So harmless to the patient.
- MRI can generate cross sectional image in any plane like axial. Sagital, coronal or any degree between without moving the patient.
- It eliminates the need for biopsy or exploratory surgery.
- Contrast agents have no side effects.
- Used for radiation therapy simulation.
- It is the best diagnostic technique for evaluating muscle and tendon disorders.

Disadvantages

- Movement is restricted by using anesthesia because examination can take 20-90 minutes or more.
- Orthopedic hardware (plates, screw, nail, pins and artificial joints) or any implant in the area of scan leads to artifact on the image.
- Noise during scan.
- Requires room without any metallic structure.
- Needs expertization for scan and interpretation.

- Because of the strong magnets, certain metallic objects objects like metal chain or other metallic objects are not allowed in the scanning room.
- It is very expensive.

Specialized MRI scans

1. **Functional MRI (fMRI):** It measures the hemodynamic response related to neural activity in the brain and spinal cord. It is best for neuro-imaging.

2. **Magnetic resonance angiography:** It is used to generate pictures of the arteries to evaluate stenosis or aneurysms. Magnetic resonance venography (MRV) is used for diagnosis of abnormalities of veins.

3. **Magnetic resonance spectroscopy (MRS):** It provides the chemical information of that region. MRS detects the intracellular relationship of brain metabolites.

4. **Interventional MRI:** Images produced by MRI are used to guide minimally invasive procedures.

5. **Real time MRI:** It refers to continuous monitoring (filming) of moving objects in real time

6. **Diffusion MRI:** It measures the diffusion of water molecules in biological tissues.
   - In biological tissues the diffusion of water is anisotropic e.g. a molecule inside the axon move principally along the axis of neural fibers.
   - In **diffusion tensor imaging (DTI)** diffusion is measured in multiple directions.

ECHOCARDIOGRAPHY

Synonyms: Cardiac echo or cardiac ultrasound.

Definition: An echocardiogram is the sonogram of the heart.

- It is a noninvasive method which produces two dimensional image of the heart.
- It provides information including shape and size of the heart, its pumping capacity, location and extent of damage to tissue etc.
- Doppler ultrasound (pulsed or continuous) not only creates 2D picture of heart and great vessels but also produces accurate assessment of the velocity of blood and cardiac tissue in any arbitrary point.
- The latest ultrasound system employs 3D real time imaging.

Indications: Used to diagnose the anatomic and functional disorder of the heart.

- Assessment of cardiac valves and their functions like valvular insufficiency and regurgitation.

- Abnormal communication between left and right side of the heart.
- Calculation of cardiac output and ejection fraction.
- Measurement of cardiac dimensions like luminal diameter, septal thickness etc.
- Intravenous contrast enhanced ultrasound (gas filled microbubbles).
- Cardiac diseases like hypertrophic cardiomyopathy, CHF, endocarditis, pericardial diseases, cardiac neoplasms and cor pulmonale etc.

Echocardiography can be performed through

1. **Trans-thoracic echocardiogram (TTE):** Standard echocardiogram is performed through transthoracic approach. The transducer/probe is placed on the chest wall and images are taken.

2. **Trans-esophageal echocardiogram (TOE):** TOE is the alternative method to perform echocardiography. A specialized probe containing ultrasound transducer is passed in to the esophagus and images of the heart are recorded.

Types of echocardiography

1. **Two dimensional echocardiography (Real time B mode):** Commonly used to examine the heart, blood vessels and offers substantial anatomic information but unable to detect abnormal blood flow, congenital heart diseases, pericardial effusions, valvular dysfunction and other functional cardiac abnormality.

2. **M-Mode echocardiography:** It is valuable for quantifying heart size, left ventricular function and imaging rapidly moving structures.

3. **Three dimensional (3-D) echocardiography:** 3 D echocardiogram of the heart viewed from the apex. It is done by using an ultrasound probe with an array of transducers and appropriate processing system.

4. **Doppler echocardiography:** In Doppler echocardiography, ultrasound is directed towards the heart and is reflected by R.B.Cs. The frequency of reflected signals is altered in relation to the velocity of RBCs. Direction and speed of the moving target can be calculated by the change in frequency of the Doppler signal. It is a sensitive technique for the detection of abnormal blood flow, valvular regurgitation etc.

S.No.	2D and M-Mode imaging	Doppler echocardiography
1.	The best image is obtained when (i) The ultrasound beam is directed at right angles to the reflective surface of the heart. (ii) By using high frequency transducers which have high resolution.	(i) The ultrasound beam is directed parallel to the moving target *i.e.* parallel to blood flow. (ii) Low frequency transducers are able to record higher velocities.

- Type of probe used: Sector scanner is more useful than linear scanner.
- Placement of probe: The probe is placed on the latero-ventral aspect of the thoracic wall, adjacent to the sternum and caudal to axial. Cardiac images are best obtained from the right side of the thorax.
- Examination table: Cut out examination table is used.

SCINTIGRAPHY

The term scintigraphy refers to formation of a two dimensional image from light flashes as a result of the interaction of energy with an absorbing material (scintillation). In scintigraphy, radio-isotopes are taken internally and the emitted radiation is captured by gamma camera to form a two dimensional image. There are four basic elements involved in scintigraphy.

1. The radiopharmaceutical agent (radioisotope)
2. The patient
3. Radio detector
4. Methods of processing and storing data.

Radioisotopes are unstable form of an element that gives off energy as rays (X-rays or gamma rays) or particles (beta or alpha) in the process of reaching a neutral energy state.

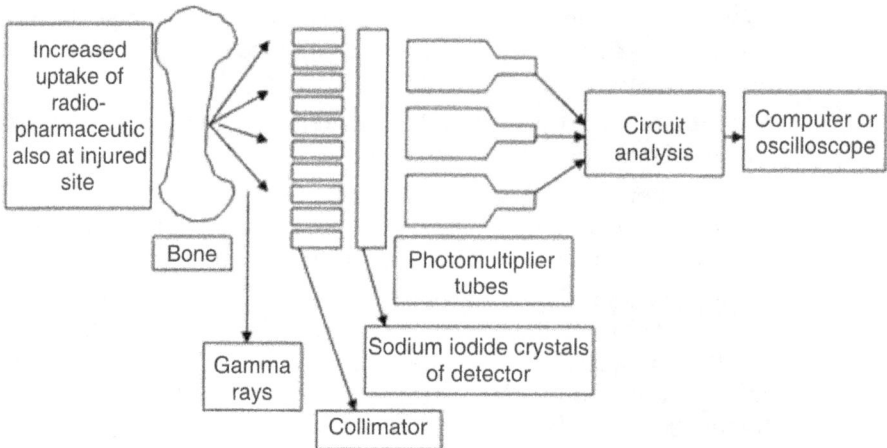

Fig. 18.1: Components of scintigraphic system

- After injection of technetium-99m labelled diphosphonate, there is uptake of radiopharmaceutic in the skeleton which is proportional to metabolic activity and blood flow.
- Gamma rays that reach the crystal of detector give off a flash of light called an event.
- This light reached to each photomultiplier tube is translated to electrical signal.

- Analysis of these signals results in localization of the event and determination of the energy of the incident gamma rays.
- This information is used to form the image, either on a oscilloscope or within computer memory.

1. Technetium-99m (99mTc) is the most commonly used isotope for scintigraphy.
2. 99mTc labelled methylene diphosphonates (Tc99m-MDP): For assessment of musculoskeletal system.
3. 99mTc labelled macroaggregates of albumin (Tc99m-MAA): For assessment pulmonary function.
4. 99mTc labelled autologous whole blood cells (Tc99m-MDP): For assessment of sites of occult infection.

Indications: To identify and describe pathologic information that can not be discerned by other imaging methods.

1. **Skeletal scintigraphy**
   - For screening of occult lameness.
   - For determining of activity of a bone lesion.
   - The distribution of Tc 99m-MDP is dependent on delivery (blood flow) and uptake (osteoblastic activity).
   - Sensitive method of identifying hyperemia, ischemia and other abnormal pattern of flow.
     - Decreased blood flow: Frostbite
     - Increased blood flow: Acute active lesion with hyperemia or neovascularization.
   - Area of bone injury demonstrate increased rate of bone activity which subsequently results in increased uptake of radioactivity visible as hot spots on the scan.
   - More intense uptake occurs at the site of high bone activity such as fractures, tumors or infection (osteomyelitis).
   - Bone lesions typically exhibit increased activity within 24-72 hrs after injury, although may not demonstrate lesions for upto 2 weeks.
   - For diagnosis of avulsion of suspensory ligaments.
   - For diagnosis of degenerative changes of joints..

2. **Pulmonary function**
   - Like ciliary function, pulmonary blood flow, ventilation and alveolar membrane permeability, Chronic pulmonary obstructive disease (COPD).

3. **Cholescintigraphy or hepatobiliary Imino-diacetic acid scan (HIDA scan)**
   - Normal gall bladder is visualized within 1 hr of injection of radioisotope. More than 4 hr indicates cholecystitis or cystic duct obstruction.

- To diagnose obstruction of bile ducts by gall stone (cholelithiasis), tumor etc.

4. **Heart**

- Thallium stress test is a form of scintigraphy, in which thallium-201 is detected in the cardiac tissues.

GAMMA CAMERA

Gamma camera is the most common radiation detector in veterinary use. It consists of a rare earth activated sodium iodide crystal that is tightly sealed to an array of many closely aligned "light pipes". The 7 mm thick sodium iodide crystals gives off a flash of light when struck by gamma rays or X-rays with an intensity proportional to the energy of the incident ray. Such a light flash is known as event. The light from an event reaches several of the light pipes which are made up of a photomultipler tube and preamplifier. The light pipes serve to translate the light flash in to electric signals for further processing; the amount of the light reaching a light pipe is translated in to given amplitude of signals. The camera is capable of counting 10,000 to 100,000 events / second with a resolution of 2 to 3 mm.

The computer divides the surface area of the gamma camera in to a checker board matrix (128 x 128 or 256 x 256). As an event is recorded, it is assigned to a bin based on its X and Y coordinates. These bins are actually binary switches which can record up to 2^8 or 2^{16} event at a location. At the end the information is stored as a matrix of numbers (digital information) that can be displayed or analyzed in a variety of ways.

The first gamma camera was developed by Hal Anger in 1957 and named as Anger camera.

Gamma camera is used in different imaging techniques like:

1. Scintigraphy
2. SPECT (Single photon emission computed tomography).
3. PET (Positron emission tomography)

Reference: Kraft, S.L. and Roberts, G.D. (2001). Modern diagnostic imaging. *The Veterinary Clinics of North America, Equine Practice,* **17(1):** 63-94 and 115-130.

XERORADIOGRAPHY

- Discovered by Chester F. Carlson (1937).
- In xeroradiography, a latent electrostatic pattern is produced on the surface of the photoconductor (amorphous selenium).
- Xeroradiography is a type of radiography in which a picture of the body parts is recorded on paper rather than on film. It is rapid method of recording a roentgen image by a dry process.

Technique

- A plate of amorphous selenium which rests on a thin layer of aluminum oxide is charged uniformly by passing it in front of scorotron.
- X-ray photon interacts with amorphous coat of selenium, charge diffuse out in proportion to energy of the X-ray (photoconduction).
- The charge distribution on plate attracts and holds a fine powder (toner) which can be transferred to a paper surface.

Difference from radiography: Image produced differ from the conventional radiography in that

- Margins between tissues of varying densities are more clearly defined, very wide exposure latitude, increased image resolution, no requirement of dark room, recording of image on opaque paper thus no need of illuminator.
- The property of edge enhancement in xeroradiography allows good delineation of structures.

Disadvantages: Xeroradiography cannot be used for very thick parts because very high exposure is needed.

IMAGE INTENSIFIER

X-ray image intensifier converts low intensity X-rays in to visible image. The device contains-

- (*i*)   Input fluorescent screen (zinc cadmium sulphide crystals or cesium iodide).
- (*ii*)   Photocathode (cesium and antimony).
- (*iii*) Electron optics/ electrostatic focusing lenses.
- (*iv*)   Accelerating anode.
- (*v*)   Output fluorescent screen and output window (silver activated zinc cadmium sulphide).
- (*vi*)   Aluminum filter: It allows high energy photoelectrons to pass through towards the output phosphor but prevents retrograde movements.

These parts are mounted in a high vacuum environment within glass or metal or ceramic lead glass window over the output phosphor.

Components

- C-arm (encompasses the actual x-ray source and image intensifier).
- Table
- Fluoroscopic exposure
- Post processing software
- Viewing monitor

Procedure and mechanism

1. X-ray beam exists from the body part enters the intensifying tube and interacts with input fluorescent screen.
2. X-ray energy is converted to visible light.
3. The visible light photons strikes the photocathode and photoelectrons are emitted (photoelectron emission is directly proportional to intensity of light falling on photocathode).
4. These photoelectrons accelerate towards anode (due to potential difference between photocathode and anode) and interact with output phosphor. Fifty five to seventy five times light photons are emitted.
5. The image can be viewed through a mirror optical system that magnifies the image through mirrors and lenses. Brightness and contrast can be controlled by television monitoring system. The motion of an organ can be recorded by video recording attachment. A spot film camera can also be attached to the image intensifier.

Advantages

-   Image produced is 1,000-5,000 times brighter than fluoroscopy.
-   Lower mA is required so less radiation exposure to the patient.
-   Examination can be done in lighted room.
-   Brightness and contrast can be controlled by television monitoring system.

SUBSTRACTION TECHNIQUE

The substraction technique is a photographic method for eliminating certain unwanted shadows from a roentgenographic film. The method was first described by Zieded des Plantes (1935).

Uses: When the contrast medium in the vascular system is in-evitably obscured by superimposed bone shadows. The bone shadows from an angiogram are deleted to clearly visualize the pattern of blood supply.

Procedure

-   Take a survey radiograph and an angiogram exactly in the same plane and without motion.
-   Make a negative of the survey radiograph (substraction mask) i.e. black.
-   When mask is superimposed over the angiogram, bony structures showed high density and vessels are clearly visible.
-   Illuminator is required.

References

1. Vezina, J.L. and McRae, D.L. (1962). Simplified method of substraction radiography. J. Canadian Radiologists, 13: 123-125.

2. Horenstein, R., Lundh, A. and Sjogren, S.E. (1964). Substraction method. Acta Radiologica (Diag.), 2: 264-272.

POSITRON EMISSION TOMOGRAPHY (PET)

- PET is a nuclear medicine imaging technique.
- It produces a 3 D image of functional processes in the body.
- A positron emitting radionuclide is introduced into the body on a biologically active molecule. Such labelled compounds are known as radiotracers.
- The gamma rays emitted by positron emitting radionuclide are detected by the system.
- Images of tracer concentrations in 3 D or 4 D within the body are then reconstructed by computer analysis.
- The biologically active molecule used is *Fluoro Deoxy Glucose (FDG)*. The concentrations of the tracer imaged then give tissue metabolic activity, in term of regional glucose uptake.
- Total dose of radiation in PET scanning is 5-7 m Sv.
- In combined PET/CT scan is 23-26 m Sv for a 70 kg patient.

Method

- The short lived radionuclide tracer isotope is chemically incorporated in to a biologically active molecule and injected in the body.
- After a waiting period (For FDG-1 hr.) the patient is placed in imaging scanner.
- It emits positron (gamma rays). These rays reach to scintillator, creating a burst of light which is detected by photomultiplier tubes or silicon avalanche photodiodes (Si APD).

Commonly used radionuclide in PET scanning

S.No.	Radionuclide	Half life (in Minutes)	Remark
1.	Carbon-11	20	
2.	Nitrogen 13	10	
3.	Oxygen 15	2	
4.	Fluorine 18	110	Most commonly used as FDG
5.	Rubidium 82		Used for myocardial perfusion studies

Limitations: High cost of cyclotrons needed to produce the short lived radionuclide.

Uses

- In clinical oncology (Imaging of tumors and its metastases).
- PET and SPECT (single positron emission computed tomography) are capable of detecting areas of molecular biology detail even prior to anatomic change.

PET with CT or MRI: The combination of PET with CT or MRI gives both metabolic and anatomic information so that areas of abnormality on the PET imaging can be more precisely correlated with anatomy on the CT images.

AUTORADIOGRAPHY

In radiography, the part to be examined is placed between source of radiation and film. But in autoradiography, the specimen itself is the source of radiation which originates from radioactive material incorporated into it. The image is recorded on photographic emulsion.

In autoradiography, the radioactive material within the cell is located by exposure of cells to a photographic emulsion forming a pattern on the film.

FLUOROSCOPY

Fluoroscopy is the dynamic radiological study of the body parts. It is used to obtain real time moving images of the internal structures of the body through the use of fluoroscope instead of a film. Thomas A Edison (1896) discovered calcium tungstate screen and fluoroscope.

Fluoroscopic unit had three parts

1. **X-ray tube:** This tube is used for both conventional radiography and fluoroscopy. The X-ray tube is placed below the table top so as to reduce the patient input dosages. The fluorescent screen is attached on the top of table. The tube is mobile along the length of the table. The tube current is typically 0.1mA to 6mA.

2. **Fluoroscopic table:** This table is equipped with a motor driven tilting machine. The table top is made up of a radiolucent material.

3. **Fluoroscopic screen:** It is just like intensifying screen except that it contains different phosphor *e.g.* silver activated zinc cadmium sulphide. It is slightly thicker (0.45-0.65mm) than intensifying screen and emits yellow green light.

   - Modern fluoroscopes are coupled with X-ray image intensifier and CCD video camera allowing the image to be recorded and displayed on a monitor.

   - Digitalization of the image and use of flat panel detectors reduced the radiation dose.

Use: Clinical evaluation of the dynamics of the body e.g.

1. Peristalsis
2. Movements of the joints

Safety precautions

- The radiologist and his/her staff should wear radiation protection devices like lead apron, lead gloves and lead goggle etc.
- The smallest possible field should be examined with limited time.
- The fluoroscopy of head region should be avoided.
- Aluminum filter of 3mm thickness should be used at the aperture of the tube so as to reduce exposure dose of the patient. Exposure rate at the table top must not exceed 10 roentgen / minute.
- The kV should be increased so as to reduce the mA while producing a brighter image.
- Fluoroscopy should never be used as a substitute for non motion radiographic examination because of possibility of radiation hazard even after adopting all preventive measures.

Principles of Radiation Therapy, Radioisotopes and their Uses in Diagnosis and Therapy

RADIATION THERAPY

Treatment of diseases or solid tumors and occasionally that of benign conditions with the use of ionizing radiation. It can be used alone or in combination with surgery.

Synonyms: Radiotherapy, radiation oncology, XRT.

Objective: Eradication of tumor/disease with preservation of structure and function of normal tissue.

(a) To cure or shrink early stage cancer.

(b) To stop metastasis.

(c) To treat symptoms caused by advanced cancer.

Mechanism of action

Radiation damages the genes (DNA) of neoplastic cells so that the cells cannot grow.

Methods

1. **Total body radiation (TBI):** radiotherapy used to prepare the body to receive bone marrow transplant.

2. **Local radiation**

   (d) Malignant / cancer treatment

   (e) Non-malignant condition treatment e.g. trigeminal neuralgia, pterygium, prevention of keloid skin growth, prevention of heterotopic ossification.

Ionizing radiations used for cancer treatment

1. **High energy photons (X-rays and gamma rays):** It is the most common treatment of choice. Sources are cobalt, cesium and linear accelerator.

2. **Particle radiation (Electrons, protons, neutrons, alpha particles, beta particles):** Produced by linear accelerator. Used for tumors close to a body surface since they do not go deeply into tissues.

3. **Proton beam**: Require highly specialized equipments.

Radioisotopes: A radioactive isotope; one having an unstable nucleus and emitting characteristic radiation during its decay to a stable form. These radioisotopes are used to detect (diagnosis) and to treat the abnormality (therapy).

Radiosensitibity: Risk of damage of neoplastic cell to absorbed radiation.

Radiation can be delivered to the malignant area in 3 ways:

Mechanism of action

Direct effects

Radiation therapy works by damaging the DNA of cells which may be caused by electron, proton, neutron, photon or ion beam directly or indirectly ionizing the atoms which make up and DNA chain.

Indirect effect

Ionization of water leads to formation of free radicals like hydroxyl radicals which than damage the DNA. Most of the radiation effect is due to free radicals.

Radiosensitizers

Drugs that make cancer cells more sensitive to radiation are 5-fluorouracil (5-FU), Misonidazole, Nimorazole, Cisplatin and Cetuximab. One of the major limitations of radiotherapy is that the cells of solid tumors become deficient in oxygen. Solid tumors outgrow their blood supply causing hypoxia. Oxygen is a potent radiosensitizer increasing the effectiveness of a given dose of radiation by forming DNA damaging free radicals. Tumor cells in hypoxia condition are 2-3 times more resistant to radiation damage than normal oxygen supply.

Radioprotectors

Amifostine protects normal cells from radiation.

Tissue compensator

Tissue compensators are used to give a more homogeneous dose to the treatment site. *e.g.*, A dog's head has varying thickness' which cause the radiation dose to vary. A tissue compensator is used to fill in the difference in thickness. These compensators can be made of Plexiglas or lead and may be made to fit on trays attached to the radiation therapy machine.

Hyperthermia

Heat produced by microwaves and ultrasound improves the effect of radiation.

Linear energy transfer (LET)

The amount of energy deposited as the particles traverse a section of tissue is referred as LET. High LET particles such as carbon or neon ions may have an anti tumor effect which is less dependent of tumor oxygen because these particles act via direct damage.

- X-rays produces low LET radiation which causes only single strands break of DNA helix. It can be repaired readily.
- Neutron produces high LET radiation which causes double strand breaks of DNA helix. It cannot be easily repaired.

Dose: for solid epithelial tumors- 60-80 Gy and for lymphomas 20-40 Gy.

Principles of radiation therapy: regardless of its nature or mode of production ionizing radiation is characterized by the mechanism of energy dissipation in tissues (ionization and excitation) of atoms and molecules in the cells. Ionizing radiation deposits energy in cells via random ionization of cellular macromolecules, when the cellular DNA is damaged so that one or both DNA strands are broken, the cells are unable to continue to divide and die when attempting mitosis.

Radiation tolerance: There is a radiation dose limit at which normal tissues are irreparably damaged. This is termed as radiation tolerance. The dose of the radiation that can be delivered to a tumor is limited by the radiation tolerance of surrounding normal tissues.

Radiocurability: A radiocurable tumor is one that can be destroyed by a dose of radiation that is well tolerated by surrounding normal tissues. Tumors in decreasing order of radiocurability are:

Squamous cell carcinoma/ pappiloma > Sarcoid, soft tissue sarcoma (neurofibrosarcoma, fibrosarcoma, hemangiosarcoma) > Melanoma.

Indications: Used for the treatment of solid tumors

1. Tumors that are confined to a limited anatomical area but are locally invasive are best treated by irradiation.
2. Tumors that cannot be completely surgically removed
3. Tumors that have already recurred after surgical removal.

Prognosis: depends on tumor volume, type and location.

Radiation therapy before surgery: Used to improve respectability by decreasing tumor size and sterilizing tumor margins.

Radiation therapy after surgery: Used to eradicate tumor cells left behind after an incomplete excision.

Effect of radiation on different types of cancer: The response of tumor to radiation is described by its radio sensitibity.

1. Tolerant tissue to radiation or radioresistant	Connective, muscular and osseous, nerve tissue tumor, melanoma
2. Most vulnerable to radiation Or Highly radio-sensitive cancer	Hematopoietic (leukemias), lymphoid tissue and germ cell tumors
3. Intermediate sensitivity to radiation or Moderately	radiosensitive Squamous epithelium and glandular tissue tumor. Require higher dose of radiation (60-70 Gy)

Commonly used radiation sources (radionuclides) for brachytherapy:

S.No.	Radionuclide	Type	Half life	Energy
1.	Cesium-137	γ-rays	30.17 year	0.662 MeV
2.	Cobalt-60	γ-rays	05.26 years	1.17 MeV
3.	Iridium-192	γ-rays	74 days	0.38 MeV
4.	Iodine-125	X-rays	59.6 days	31.4 KeV
5.	Palladium-103	X-rays	17 days	21 KeV
6.	Ruthenium-106	β- particles	1.02 years	3.54 MeV

Side effects: Side effects are usually limited to the area of the patient's body that is under treatment. Severity and longevity of side effects depends on organ that receives the radiation and treatment itself (type of radiation, dose fractions and concurrent chemotherapy).

Acute side effects: Side affects during treatment due to higher doses and depends on the area being treated.

(a) **Damage to the epithelial surface:** Moist desquamation in dogs (cats more commonly get dry desquamation with dry flaky skin and itchiness) Licking worsens the condition. Radiation burns is followed by crusts formation and then the skin heals under the crusts. The whole process takes about 2-3 weeks.

(b) **Mucositis** is seen when the gums, tongue, cheeks, throat or other mucus membrane-lined tissue is in the treatment field (often with nasal or oral tumours). The mucosa will get very red, ulcerate or blister. Halitosis, drooling, and difficulty eating can occur. Some cats and small dogs could require a temporary feeding tube if a large portion of their mouths are in the treatment field.

(c) **Ocular side effects** are of concern if the eyes are in the treatment field. Acute side effects include dry eye and corneal irritation and corneal ulcer formation.

(d) **Swelling and edema:** May cause obstruction of the lumen like trachea, bronchus. Steroid therapy during radiotherapy reduces swelling.

(e) **Infertility:** The gonads are very sensitive to radiation.

(f) Anemia

Late side effects: Occurs months to years after treatment and are due to damage of blood vessels and connective tissue cells. Late effects can be reduced by fractionating treatment into smaller parts.

1. Fibrosis and thickening of the skin. Delayed wound healing in the area that was treated.

2. Epilation (loss of hair): Permanent epilation occurs by a single dose of 10 Gy. Permanent change in hair coat

3. Dryness like dry mouth (Xerostomia), loss of taste and dry eyes (xerophthalmia).

4. Late side effects of radiation (permanent keratoconjunctivitis sicca or dry eye, cataracts, and retinal degeneration can be irreversible side effects of radiation therapy if the eye gets full dose or even scatter radiation.

5. Lymphedema

6. Spinal cord malacia, kidney fibrosis or scarring, lung fibrosis.

7. Cancer: radiation is a potential cause of cancer.

> • Pets must be fasted for at least 12 hours prior to their radiation therapy (due to anaesthesia needs)
>
> • There are four important factors affecting cell sensitivity to radiation that is referred as the 4 R's- *repair, repopulation, redistribution, and re-oxygenation.*

Teletherapy (External beam radiation therapy or EBRT or XRT): The source of radiation is outside the body at some distance (80-100 cm) from the target tissue. The radiation source is focused on the area affected by the cancer. Teletherapy is carried out by linear accelerator that deliver high energy X-rays or by gamma beam emitted from radioactive cobalt-60 source.

(a) *Conventional external beam radiotherapy (2DXRT):* Radiation is delivered via two dimensional beams using linear accelerator machines. 2DXRT consist of a single beam of radiation delivered to the patient from several directions.

(b) *Stereo tactic radiation:* It uses focused radiation beams targeting a well defined tumor using imaging scans.

Treatment planning:

1. *Simulation/marking session:* Marking of radiation field/treatment port. Imaging techniques like CT scan can be used to check the size of the tumor.

2. *Dose (In gray-Gy):* The total dose is divided into several small fractions and given for weeks (5 days a week for about 5-8 weeks). This allows least damage to the normal tissue. In veterinary medicine, a common total dose would be 48-60 Gray given over 12-20 fractions in 4 weeks.

    (i) *Hyper-fractioned radiation:* When the daily dose is divided twice a day without changing the duration of the treatment.

    (ii) *Accelerated radiation:* When the total dose of radiation is given over a shorter period of time.

Indications: High energy treatment units is used for deep seated tumors like bone tumors, Intracranial or intranasal tumors or large radiosensitive tumors (localized lymphomas), oral cavity tumors. Low energy radiation (orthovoltage radiation) can be used for treating cutaneous tumors.

1. Kilovoltage (Superficial) X-rays are used for treating skin cancer and superficial tumors.

2. Megavoltage (deep) X-rays are used to treat deep tumors (brain, lung and bowel).

    Diagnostic X-rays- 20-150 KV (kilovolts)

    Superficial X-rays - 50-200 KV

    Orthovoltage X-rays- 200-500 KV

    Supervoltage X-rays- 500-1000 KV

    Megavoltage X-rays- 1-25 MV (megavolts).

- Linear accelerator produce megavoltage X-rays.

Brachytherapy (Internal beam radiation therapy or sealed source radiotherapy or short distance therapy):

Sealed radioactive substances (gamma or beta emmiter) are placed in to the tumor or into the body cavity close to the tumor. Advantage of brachytherapy is that it delivers a high dose of radiation to a small area e.g.

- I -125 and I -131 for thyroid cancer or thyrotoxicosis
- Iridium-192 in brain cancer.

Permanent brachytherapy (long term)

Permanent implant is left in place forever and slowly delivers its dose of radiation until the radioactive source has decayed to a negligible level. This therapy is generally not recommended due to potential radiation hazard. Rice grain size pellets or seeds are placed in to the tumor using thin hollow needles.

Temporary brachytherapy (short term)

Radiation source is left in the tumor bed for a specified time and is removed when the prescribed radiation dose has been given. There is no residual radioactivity once the implant is removed. Hollow stainless steel needles, tubes or fluid filled balloons are placed in to the area of treatment. The temporary sources are usually placed by a technique called afterloading *i.e.* a hollow tube or applicator is placed surgically in the organ to be treated and the sources are loaded in to the applicator after the applicator is implanted. This minimizes the radiation exposure to the health care personnel.

(i) *High dose rate (HDR):* Radiation source is put into place for about 10-20 minutes at a time and then removed. Repeat twice a day. Remote afterloading, a new technique of radiation sources has been developed to eliminate radiation hazards and shorten treatment time.

Advantages: More target specific (high dose of radiation can be delivered directly to the tumor), less exposure of radiation to the body or adjacent normal tissue and cost effective.

Side effects: Nausea, burning sweating

(*ii*)   Low dose rate (LDR): @ 0.3-0.5Gy/hr. The radiation source is placed for up to 7 days.

Types of brachytherapy

1.   **Interstitial brachytherapy (IB) or Curie therapy:** Radiation source is implanted directly into the tumor in the form of small pellets, grains, seeds, wires, tubes or containers under local or general anesthesia. It is an ideal technique for irradiation in horses. Treatment has its maximal effect in the early post-operative period (2-3 weeks) when the tumor burden is minimal. Too early irradiation is not recommended because of decreased wound healing. Not used for bone tumors.

Indications

- Operable cutaneous tumors with indistinct margins.
- Limited to soft tissue tumors like squamous cell carcinoma, sarcoid etc.

2.   **Intracavitary radiation:** Radiation source is placed into the body cavity such as vagina, uterus, rectum, chest etc.

3.   **Plesiotherapy (Surface therapy):** radiation source is applied on to the tumor surface. Most commonly used radiation source is Strontium-90. A Strontium-90 applicator is used for irradiation (half life 30 year). It has shallow depth of penetration (60% radiation dose is absorbed within the first mm of tissue).

     Indications: Small superficial (≤ 1 mm) tumors of eyelid, conjunctiva and corneoscleral tissues including squamous cell carcinoma and melanoma.

4.   **Intraluminal brachytherapy:** Radiation source is placed in to the lumen such as trachea, esophagus etc.

RADIOPHARMACEUTICAL

These are the drugs containing radioactive substances. These drugs may be placed in to the vein, body cavity or can be taken by mouth. According to the type of radionuclide, the tracer will collect in one or more areas of the body. Since the tracer emits radiation, it is easily tracked by Geiger counter (a device that measures radioactive levels) or scanning device.

Radiopharmaceuticals were first used in medicine for diagnosis during early 1930s. Therapy by radiopharmaceuticals is known as radiopharmaceutical therapy or unsealed source radiotherapy.

- Gamma scintillation camera was invented by Hal Anger (1950) which is helpful for diagnosis and treating diseases.

- Positron emission tomography (PET scan) was the first diagnostic tool invented by Peter Alfred Wolf that used radioisotope (Fluorine-18) in medicine.
- Edith Quimby was an American radiologist who first accurately measured the amount of radiation necessary to allow body traces.

Isotopes: The element that have same atomic no. but different atomic mass and physical properties.

Radioisotopes: Isotopes that have unstable no. of protons and neutrons. Cyclotrons are used to manufacture proton rich radioactive isotopes.

1. *Strontium 89 and samarium-153*: Used for diagnosis of bone tumors or bone metastasis. These substances accumulate in the area of bone tumor. Radiation produced by these substances kills the cancer cells. Spread of cancer to more than one bone can be best treated by these drugs.

2. *Iodine- 131 or radioactive iodine or radioiodine*: It is the first radioisotope used in medicine for thyroid cancer and thyrotoxicosis. It leaves the body within a few weeks mainly through the urine but also through saliva, sweat and stool. Half life is 8.0197 days.

3. *I - 131 Metaiodobenzylguanidine (I^{131} - MIBG)*: It is used for imaging and treating Pheochromocytoma and neuroblastoma.

4. *Phosphorous- 32 or chromic phosphate P 32*: It is used for treating brain tumors. Also used to control polycythemia vera where an excess of RBCs are produced by bone marrow.

5. *Cr- 151 or Chromium (sodium chromate)*: It attaches strongly to the hemoglobin of RBCs. It is an excellent isotope for determining the flow of blood through the heart and to determine the age of RBCs for diagnosis of anemia.

6. *Cobalt-59 or Cobalt-60* is used to study defects in vitamin B_{12} absorption.

7. *Technetium-99*: It is the most commonly used isotope in nuclear medicine.

8. *Rhenium-186*: it is used for pain therapy.

9. *Thallium-201* and *Technetium-99* are used in myocardial perfusion imaging for detection and prognosis of coronary diseases.

10. *Radio-labelled antibodies*: Monoclonal antibodies attack only a specific molecular target on certain cancer cells e.g. Ibritumomab tiuxetan, an anti CD monoclonal antibody conjugate to Yttrium-90 used for radio immunotherapy.

SPECIAL RADIATION THERAPY

1. **Three dimensional conformal radiation therapies (3-D-CRT):** MRI, CT or PET scan locates the cancer in 3D. The radiation beams are matched to the shape of the tumor and delivered to the tumor from several directions.

2. **Intensity modulated radiation therapy (IMRT):** It is an advanced form of teletherapy. MRI, CT or PET scan locates the cancer in 3D. In IMRT along with aiming photon beams from several directions, the intensity (strength) of the beams can be adjusted.

3. **Conformal proton beam radiation therapy:** In this therapy proton beam is used instead of X-rays. Protons are generated by cyclotron or synchrotron.

4. **Stereotactic radiation therapy (SRT)/ stereotactic radiosurgery:** SRT delivers a large, precise dose of radiation to a small tumor area. Brain tumors are treated with this technique.

5. **Intra-operative radiation therapy (IORT):** IORT is the delivery of the radiation to the cancer during surgery or immediately after surgical removal of cancer. Radiation may be given internally or externally. IORT is useful for abdominal or pelvic cancers that cannot be completely removed by surgery.

6. **Image guided radiation therapy (IGRT) or 4D radiotherapy:** Real time imaging combined with real time adjustment of the therapeutic beams is used for IGRT.

7. **Fast neutron therapy:** High energy neutrons of >20 MeV are used to treat cancer.

8. **Selective internal radiation therapy (SIRT):** It is a form of radiation therapy which is used to treat those cancers that cannot be treated surgically e.g. hepatic cell carcinoma or metastasis of liver. The treatment involves injecting tiny microspheres (made of Yttrium-90 covered by glass or resin) of radioactive material into the arteries that supply the tumor. The microspheres embolize and stop the blood supply to the tumor. Yttrium-90 emitted beat particles (half life-2.6 days).

9. **Intra-operative electron radiation therapy / precision radiotherapy (IOERT):** IOERT is the application of electron radiation directly to the residual tumor or tumor bed during cancer surgery.

Radiation therapy is typically applied to one site in an attempt to provide loco-regional therapy (primary tumour and regional lymph nodes) of a cancer but is not efficacious against metastatic cancer

---

20 Radiographic Signs of Diseases of Different Body Systems

DIGESTIVE SYSTEM

1. **Vascular ring anomaly**
    *   Positive contrast esophagogram shows dilation of the esophagus cranial to partial esophageal obstruction over the heart base.
2. **Esophageal foreign body**
    *   Radio-dense foreign bodies can be seen on survey radiograph.
    *   Radiolucent foreign bodies require positive contrast radiography.
3. **Esophageal stricture**
    *   Positive contrast esophagogram shows dilation of the esophagus cranial to stricture.
    *   Esophagography is the best method to determine the number, location and length of stricture.

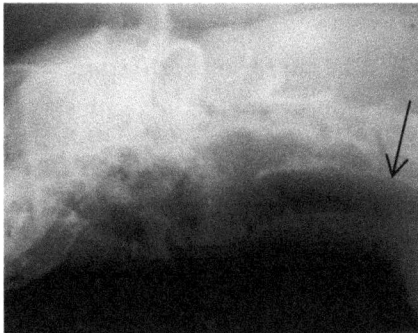

Fig. 20.1: Esophageal stricture in a dog

4. **Esophageal diverticulum:**
    *   Survey radiograph demonstrates air or ingesta filled dilation of esophagus.
    *   Positive contrast esophagography is advantageous to delineate the fistula.

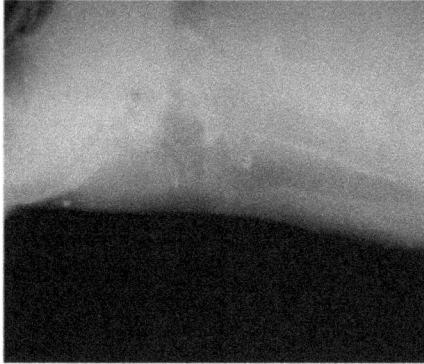

Fig. 20.2: Esophageal diverticulum in a dog

5. **Esophageal fistula**

   - Survey radiograph of the thorax demonstrates an alveolar, bronchial and/or interstitial lung pattern in the affected lobe.

   - Confirmatory diagnosis is made by Positive contrast esophagography using water soluble iodine preparations.

   - Barium sulphate is contraindicated.

6. **Crico-pharyngeal achalasia**

   - Survey radiograph of pharynx is normal.

   - Signs of aspiration pneumonia on thoracic radiograph.

   - Differential diagnosis is made by fluoroscopy during a barium swallow.

7. **Acute gastric dilation / Gastric dilation with volvulus (Bloat)**

   - Right lateral abdominal radiograph shows gas distended gastric shadow with various degree of gas filled small or large intestine.

   - Volvulus is suspected if a tissue density separates the gas filled gastric shadow in to two chambers (Double-bubble) or pylorus is dorso-cranial to the fundus.

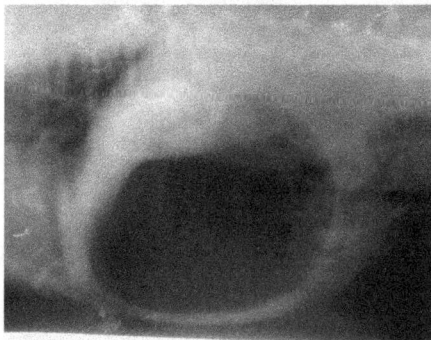

Fig. 20.3: GDVS in a dog (Curtesy: Dr. Ramesh Tiwary)

8. **Gastric perforation**
   - Free peritoneal gas present or serosal detail is obscured.
   - Contrast radiography using water soluble iodine preparations reveals leakage of contrast material into the peritoneal cavity.

9. **Gastric foreign body**
   - Radio-dense foreign bodies can be seen on survey radiograph.
   - Radiolucent foreign bodies can be outlined by positive contrast radiography (1-2 ml/kg b. wt.).

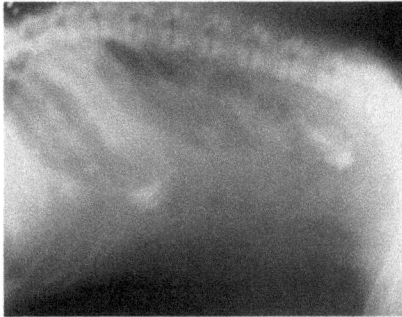

Fig. 20.4: Foreign bodies in stomach and intestine in a dog

10. **Gastric retention and outflow obstruction**
    - Static contrast images are more diagnostic than non-contrast enhanced radiograph.
    - Gastric distension and retained barium meal for >15 hours.
    - Pyloric thickening. The **"beak"** sign or an **"apple core"** appearance on contrast radiography.

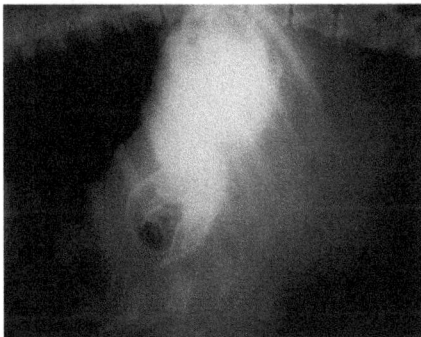

Fig. 20.5: Outflow obstruction in stomach in a dog

11. **Gastric neoplasia**
    - Abdominal radiograph shows wall thickening, mass effects, inflow / outflow obstruction or motility disturbances.

12. **Perforations associated with gastric abscess**
    - Focal increased soft tissue density associated with a portion of gastric wall.
    - Generalized loss of serosal detail and free peritoneal gas may indicate peritonitis secondary to perforation.
    - Contrast radiography using water soluble iodine preparations reveals leakage of contrast material into the peritoneal cavity.

13. **Gastro-esophageal intussusception**
    - Enlarged esophagus or caudal thoracic (epiphrenic) mass effect.
    - Contrast radiography shows gastric rugal folds within the esophageal lumen.

14. **Pyloric stenosis:** Contrast study showed gastric distension and delayed emptying.

15. **Intestinal obstruction**
    - Presence of multiple loops of gas dilated small intestine of varying diameters.
    - Small intestine (SI) diameter and the height of the 5th lumbar vertebral body at its narrowest point (L5) ratio (SI: L5) of 1.6 is considered the upper limit of normalcy.
    - Contrast radiography is required for definitive diagnosis.
    - Most proximal intestinal obstructions are evident within 6 hours of administration of barium sulphate. However up to 24 hours may be necessary to highlight distal obstructions.

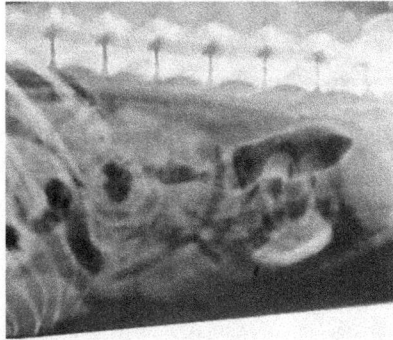

Fig. 20.6: Intestinal obstruction in a dog

16. **Linear foreign bodies of intestine**
    - On plain radiograph, small intestine is plicated and is gathered in the cranial to mid-ventral abdomen instead of being dispersed uniformly throughout.
    - Gas collects in small, eccentrically located intraluminal bubbles instead of normal curvilinear columns.

- On contrast radiography, foreign body becomes radiolucent. After contrast agent passes in to the colon, the foreign body retains barium, making it more apparent.

17. **Intussusception**
    - A mass effect and accumulation of gas and fluid in the bowel loops proximal to Intussusception.

18. **Mesentric volvulus**
    - Cranial lateral radiographs showed uniform and extensive gaseous distension of intestine.

19. **Cecal inversion**
    - A small fluid dense intraluminal mass in the proximal colon.

20. **Cecal impaction in dog**
    - Loss of gas filled cecal silhouette or foreign material in the cecum.

21. **Atresia ani**
    - Positive contrast radiography is useful to determine the position of the terminal rectum.

22. **Rectovaginal fistula**
    - Positive contrast radiography revealed free passage contrast agent between the rectum and vagina.

23. **Splenic torsion**
    - Presence of mid-abdominal mass.
    - Absence of normal splenic silhouette (C shaped spleen).
    - Air opacities within the splenic parenchyma.
    - If abdominal effusions are present. There is loss of abdominal detail.

ABDOMINAL CAVITY

Peritonitis: Loss of detail or ground glass appearance.

Diaphragmatic hernia

- Interruption of diaphragmatic contour.
- Pleural effusion and increased soft tissue density in thorax.
- Gas filled intestinal loops cranial to diaphragm.

SKELETAL SYSTEM

Osteomyelitis

- Sequestra as radiodense bone surrounded by a zone of lucency and sclerotic reactive bone.
- Bone lysis and irregular periosteal reaction. **(Fig. 20.8)**

Multiple myeloma

- Mottled osteolytic lesions or **"Punched out"** lesions.

Hemangiosarcoma of bone

- Extensive intramedullary osteolysis.

Osteosarcoma

- Focal osteoproliferative and osteolytic lesion with a periosteal reaction.
- Soft tissue swelling and an adjacent bone reaction. **"Sunburst appearance"**.

Bone cyst

- Osteolytic spaces filled with fluid/vascular elements.

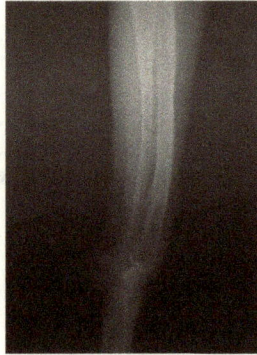

Fig. 20.7: Bone cyst

Shoulder osteochondrosis

- A radiolucent region on the caudal border of the humeral head.

Hip dysplasia

- A shallow acetabulum, flattened femoral head or both.
- Subchondral bone sclerosis.
- Periarticular osteophyte formation.
- Femoral neck remodeling (In aged canines).

Non-union

- Incomplete callus formation across a persistent fracture gap (active non-union).
- No callus formation and sclerosis of the bone ends. **"Elephant foot"** (inactive non-union) **(Fig. 20.8).**

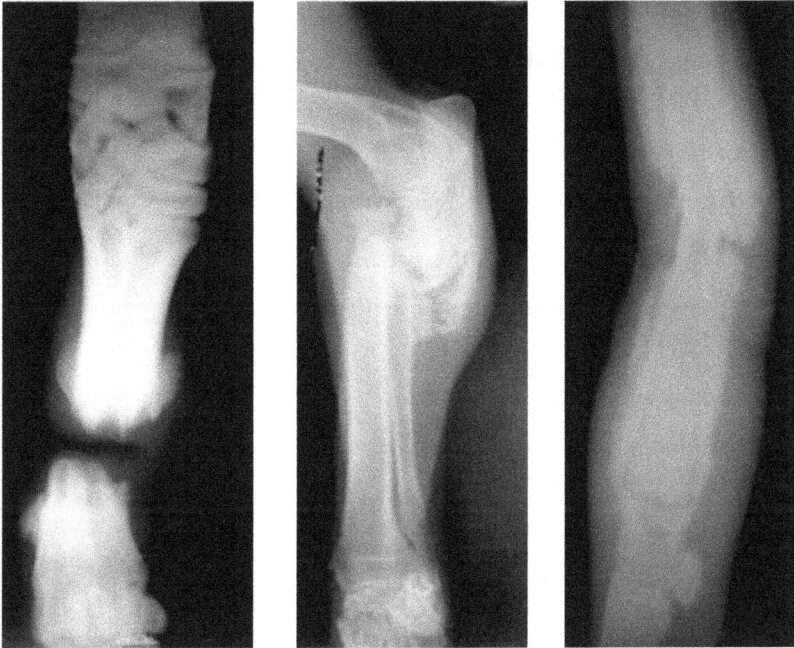

Non-union   Mal-union   Osteomyelitis

Fig. 20.8: Non-union, mal-union and osteomyelitis

Hypertrophic osteodystrophy

- Flared irregular metaphysis.
- A radiolucent line below the physes.
- Periosteal soft tissue swelling.

Osteochondromatosis

- Multiple radiopaque bone densities.

Osteochondrosis

- Flattening of subchondral bone and joint effusion.
- Osteochondral joint fragments and osteophyte formation.

Panosteitis

- Increased multifocal densities of the medullary cavity.
- Roughened endosteal surfaces of the long bones.

Legg-calve-perthes disease

- Increased joint space width due to collapse and thickening of femoral head and neck.
- Irregular density of femoral epiphyseal and metaphyseal region.

Hypertrophic osteopathy

- Long bone periosteal reaction in all four limbs and progresses into distal to proximal direction.

Bone infarction

- Irregular radiopaque density within the medullary cavity.

Nutritional secondary hyperparathyroidism

- Systemic bone resorption, fractures and increased linear metaphyseal densities.

Renal secondary hyperparathyroidism

- Demineralization, pathologic fractures and periosteal lucencies in the skull and mandible.

Rickets and Osteomalacia

- Widened irregular growth plates and "metaphyseal cupping" in young animals.
- In mature animals reduced bone density.

Hypervitaminosis-A

- Vertebral and joint fusion along with long bone exostoses.

Mucopolysaccharidosis

- Vertebral and joint fusion, pectus excavatum and bilateral coxofemoral luxation.

Degenerative joint disease (DJD)

- Periarticular swelling and joint effusion.
- Joint space narrowing due to loss of articular cartilage.
- Subchondral bone sclerosis and osteophyte production.
- In severe cases bone remodeling occurs.

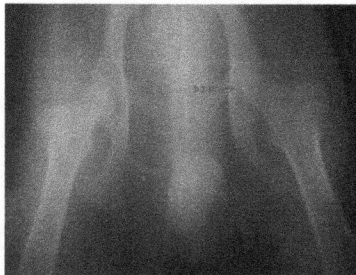

Fig. 20.9: Degenerative joint disease of hip joint in a dog

Erosive arthropathy

- Collapse of joint space, destruction of subchondral bone and periarticular swelling.

Non- erosive arthropathy

- Periarticular swelling and joint effusion only

Infectious arthritis

- Acute: Joint distension and swelling.
- Chronic (>3 weeks): Subchondral lucency and signs of DJD.

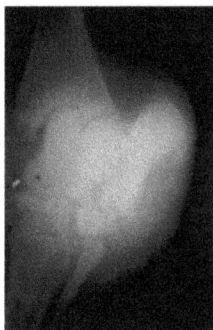

Fig. 20.10: Septic arthritis in equine

Sprain

- Stress radiography: Abnormal joint space is visible when angular forces are applied to the joint.

LAMENESS

Chronic laminitis

- Rotation of the distal phalanx which is identified by divergence of the bone in relation to the hoof wall.

Navicular disease/ podotrochleosis

- Enlarged mushroom shaped (lollipop or inverted flask shaped) vascular foramina.
- Cyst with enlarged vascular foramina in less dense region of navicular bone.
- Thinning or roughening of the flexor cortex.
- Loss of cortico-medullary junction (increased density in the medullary canal).
- Occasional fracture of navicular bone.

Pedal osteitis

- Demineralization at one or more points in the distal phalanx.

Sobchondral bone cyst of distal phalanx

- Cystic lesion within the body of the distal phalanx.

Pyramidal disease / buttress foot

- New bone growth on the extensor process of the distal phalanx / middle phalanx.

Quittor / necrosis of collateral cartilage

- The depth and dimension of the sinus can be seen on the radiograph by using the contrast radiography or by placement of a sterile probe in the tract and then take a radiograph.

Side bones

- Bone densities in the cartilages.

Keratoma of hoof

- Increased tissue density of the hoof wall and osteolysis of the distal phalanx.

Ring bones:

- In early stage no appreciable radiographic changes.
- In later stages (after 3-4 weeks), peripheral osteophytosis with articular ring bone or periostitis (periarticular new bone) with non-articular ring bone is present.

Sesamoiditis in horse

- Bony changes occur on the abaxial surface or basilar region with increased radio-dense buildups.
- Increased number and irregularity of the vascular channels.
- Increased coarseness and mottling of the bone trabeculation.

Traumatic rupture of the suspensory apparatus

- Either the proximal displacement of the intact sesamoid bone or proximal displacement of the apical portions of the fractured sesamoid bones.
- Associated swelling of the soft tissues.

RESPIRATORY SYSTEM

> - Radiograph of the thorax should be taken at full inspiration.
> - Radiograph of the abdomen should be taken at full expiration.

Sinusitis:

- Asymmetry and frontal bone erosions.

Laryngeal paralysis

- Radiograph of larynx is unremarkable.
- Radiograph of thoracic cavity revealed aspiration pneumonia and megaesophagus.

Tracheal collapse

- During inspiration distal thoracic trachea is normal.
- During expiration, distal thoracic trachea is nearly collapsed.
- Thoracic radiograph shows cardiomegaly.
- Cross sectional view / skyline view of the trachea at the thoracic inlet shows flattened rings and redundant tracheal membrane.

Lung lobe torsion

- Pleural fluid and lung consolidation.

Lung laceration

- Free air and fluid densities within the thorax.

Lung abscess

- Water density unless the abscess ruptured and drained.
- Air contrast may be seen if it connects with respiratory system.

Bronchiectasis

- Signs of atelectasis, consolidation and fibrosis.

UROGENITAL SYSTEM

Pyometra

- Homologous tubular structures of fluid density may be seen in the middle to caudal abdomen.

- **Note:** The uterus has a similar appearance in early pregnancy and immediately post-partum.

Uterine torsion

- A large air or fluid filled tubular structure can be seen on the radiograph.

Hydronephrosis

- Enlarged renal shadow.
- Excretory urography shows dilated renal pelvis.

Ureteral obstruction

- Enlarged kidney shadow.
- Excretory urography shows dilated renal pelvis with decreased renal opacification.

Cystic calculi

- Radiopaque calculi (struvite and calcium containing calculi) may be seen on the abdominal radiograph.

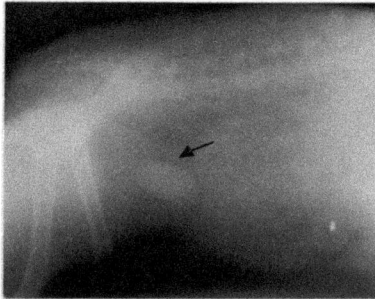

Fig. 20.11: A large calculi in the urinary bladder of a bitch

- Radiolucent calculi can be seen only after excretory urography or double contrast cystography.

Urinary bladder rupture

- Ascites and small or inapparent bladder outline.
- Positive contrast cystography reveals presence of contrast material in the peritoneal cavity.
- Negative contrast radiography should be avoided because of risk of venous air embolization.

Prostatic disease

- Cystic area within the gland.

EAR

Otitis media

- Thickening and sclerosis of the wall of one or both tympanic bullae and filling of these well aerated structures with soft tissue material.
- In advanced stage: Bony proliferation of petrous temporal bone, temporo-mandibular joint or both.

Periodontal disease

- Resorption of alveolar crest.
- Roundening of amelocemental junction.
- Widening of periodontal space, loss of lamina dura and
- Lysis of the bone surrounding the teeth.

Periapical disease

- Widening of periodontal space surrounding the apex, bone lysis or sclerosis adjacent to the apex.
- Resorption of tooth root and osteomyelitis of the adjacent bone.

CARDIOVASCULAR SYSTEM

Patent ductus arteriosus (PDA)

- Cardiomegaly (Left atrial and ventricular enlargement).
- Pulmonary vessel enlargement.
- Dilation of descending aorta.
- DV view represents the four bulges on the left side of the heart.

Pulmonic stenosis and Double chambered right ventricle

- Right ventricular enlargement / hypertrophy.
- Post-stenotic pulmonary artery enlargement.

Aortic stenosis

- Radiograph is unrewarding in the diagnosis.
- Normal cardiac silhouette, subtle enlargement of ascending aorta and left ventricular.

Persistent right aortic arch

- Positive contrast radiography by oral barium sulphate suspension reveals esophageal constriction at the base of the heart.
- Esophageal dilation cranial to the constriction.

Tetralogy of fallot

- Right ventricular enlargement.
- Main pulmonary artery dilation.

Atrial septal defect / Atrioventricular septal defect

- Right ventricular enlargement.
- Enlargement of pulmonary vessels.

Ventricular septal defect

- Left ventricular / biventricular enlargement depending on size of the defect.
- Pulmonary vessel enlargement.

Pericardial effusion

- Spherical / globoid enlargement of the cardiac silhouette with loss of normal contour of the individual chambers.
- Widening of caudal vena cava.
- Pleural effusion may be present.
- Pneumopericardiography should be done for complete evaluation of pericardial sac.

Constrictive pericarditis

- Heart appears globoid and cardiac silhouette loses its angle and waist.
- Enlargement of caudal vena cava.
- Pleural effusion may be present.
- Compression of right ventricular outflow tract.
- Fluoroscopy shows reduced motion of wall of the cardiac chambers.

Cardiac neoplasms

- Enlargement of cardiac silhouette, elevation of trachea, distension of caudal vena cava.
- Pleural effusion and pulmonary edema.

Caval syndrome / heartworm disease

- Enlarged right atrium and ventricle.
- Enlarged pulmonary trunk
- Enlarged, tortuous branches of the pulmonary artery.
- Abdominal radiograph reveals hepatomegaly and cardiomegaly.

Chylothorax

- Presence of pleural fluid.

Pyothorax

- Bilateral pleural effusion which obscures pulmonary detail and cardiac silhouette.

Mediastinal masses

- Mediastinal widening.
- Tracheal displacement to the right.
- Aortic displacement to the left.

Congenital diaphragmatic hernia (Pleuroperitoneal or peritoneo-pericardial):

- Enlargement of cardiac silhouette.
- Dorsal displacement of trachea.
- Interruption of diaphragmatic outline.
- Small intestinal gas patterns over the cardiac silhouette are pathognomonic.

Left heart failure due to dilated cardio myopathy

- Distended and tortuous pulmonary veins due to increased pulmonary vein pressure.
- Enlarged left atrial and left ventricular chamber.

Mitral regurgitation

- Left atrial enlargement with separation and collapse of main stem bronchi
- Left ventricular enlargement may be evident.
- Distension of pulmonary veins.

Tricuspid valve dysplasia

- Moderate to extremely severe cardiomegaly.

Cor triatriatum

- Radiography is unremarkable for diagnosis.
- Enlargement of caudal vena cava.

Pericardial cyst

- Globoid or odd shaped cardiac silhouette.

Pericardial rupture

- Kinked caudal vena cava cranial to diaphragm.
- Abdominal radiograph shows abdominal effusion and hepatomegaly.

Arterio-venous fistula:

- Cardiomegaly, atrial enlargement and increased lung vascularity.
- Angiography revealed tangled mass of tortuous vessels.
- Rapid filling of veins with poor distal arterial opacification is a reliable sign of a fistula.

MULTIPLE CHOICE QUESTIONS

1. All are the principles of radiation safety except
   (a) **Lower the kV**
   (b) Increase the target to skin distance
   (c) Use optimum filtration in the primary beam
   (d) None of the above

2. Roentgen is measured in
   (a) **Air**   (b) Fat
   (c) Solid tissue   (d) Soft tissue

3. A radiolucent foreign body lodged in the esophagus could be more easily seen
   (a) On a dorso-ventral projection
   (b) **Following a barium swallow**
   (c) By making exposure during expiration
   (d) None of the above

4. X-ray films should be stored in a
   (a) Hot and humid location   (b) Cool and humid location
   (c) **Cool and dry location**   (d) Hot and dry location

5.. The temperature of the filament affects the
   (a) Average energy of the X-ray produced
   (b) Speed of the produced electrons
   (c) **Quantity of the X-ray produced**
   (d) None of the above

6. Which part of the X-ray tube is the limiting factor in the maximum energy of the x-rays that can be produced
   (a) Cathode   (b) Filament
   (c) **Anode**   (d) Glass envelope

7. Increasing the atomic number of the target material
   (a) Increases life of the tube
   (b) Increases the melting point

(c)   **Increases the probability of electron interaction**

(d)   Increases the cost of the machine

8. Cystography is indicated to diagnose

(a)   **Structural abnormality of the bladder**

(b)   Atony of the bladder

(c)   Both

(d)   None

9. Which of the following statement is wrong

(a)   Intravenous pyelography should not be indicated in severely dehydrated cases

(b)   **Myelography is very useful in cases of meningitis**

(c)   Barium enema is useful to outline the colon and rectum

(d)   None

10. Barium swallow should not be used in cases of

(a)   **Oesophageal perforation**   (b)   Mucosal disease of the esophagus

(c)   Both   (d)   None

11. The function of the grid is to

(a)   Reduce developing time required (b) Eliminate the need of collimator

(c)   **Absorb secondary radiation** (d)   None

12. The objectives of the developing solution is to

(a)   Harden the emulsion and remove unreduced silver halide crystals

(b)   **Soften the emulsion and reduce exposed silver halide crystals**

(c)   Harden the emulsion and remove free silver

(d)   Soften the emulsion and reduce free silver

13. When a lead apron is worn for the radiation protection, the film badge should be worn

(a)   **Outside the apron near the neck**

(b)   Outside the apron at the waist level

(c)   Behind the apron at the waist level

(d)   Behind the apron at the chest level

14. X-rays are high energy

(a)   **electron**   (b)   Proton

(c)   Neutron   (d)   None

15. Instrument which is used to limit the primary X-ray beam to the size of the film

(a)   **Collimators**   (b)   Grid

(c)   Filter   (d)   Cassette

16. Instrument which is used for absorbing scattered radiation is
    (*a*) Collimators   (*b*) **Grid**
    (*c*) Filter   (*d*) Cassette
17. Instrument which absorb harmful and non-penetrating X-rays
    (*a*) Collimators   (*b*) Grid
    (*c*) **Filter**   (*d*) Cassette
18. Equipment used to keep an X-ray film for part being radiographed is
    (*a*) Collimators   (*b*) Grid
    (*c*) Filter   (*d*) **Cassette**
19. If film will be dark then it indicate condition of
    (*a*) **Over exposure**   (*b*) Under exposure
    (*c*) No exposure   (*d*) None
20. If there is too light X-ray and detail will be lost then it indicate
    (*a*) Over exposure   (*b*) **Under exposure**
    (*c*) No exposure   (*d*) None
21. Developing time for X-ray film is
    (*a*) **5 minutes**   (*b*) 10 minutes
    (*c*) 15 minutes   (*d*) 20 minutes
22. Which is composition of developer
    (*a*) Sodium sulphite   (*b*) Sodium carbonate
    (*c*) Potassium bromide   (*d*) **All of the above**
23. Which is composition of fixer
    (*a*) Sodium thiosulphate   (*b*) Sodium sulphite
    (*c*) Acetic acid   (*d*) **All of the above**
24. Washing time for X-ray film is
    (*a*) 5 minutes   (*b*) **30 seconds**
    (*c*) 15 Seconds   (*d*) 20 minutes
25. Fixing time for X-ray film is
    (*a*) 5 minutes   (*b*) 30 Seconds
    (*c*) **10 Minutes**   (*d*) 20 minutes
26. Final washing time for X-ray film is
    (*a*) 5 minutes   (*b*) 30 Seconds
    (*c*) 10 Minutes   (*d*) **20 minutes**
27. Weak and old developer produce defect on X-ray film is
    (*a*) Gray non contrast radiograph (*b*) Dark spot
    (*c*) White spot   (*d*) **Fog on film**

28. Spilling of a drop of developer on X-ray film during loading into cassette produce
    (a) Gray non contrast radiograph (b) **Dark spot**
    (c) White spot   (d) Fog on film

29. Spilling of a drop of fixer on X-ray film during loading into cassette produce
    (a) Gray non contrast radiograph (b) Dark spot
    (c) **White spot**   (d) Fog on film

30. Metallic base containing medicine pasted on animal body during radiography produce
    (a) Gray non contrast radiograph (b) Dark spot
    (c) **White spot**   (d) Fog on film

31. Contrast radiography of bronchus and bronchial tree is known as
    (a) **Bronchography**   (b) Myelography
    (c) Intravenous pyelography   (d) Dacrocystorhinography

32. Contrast radiography of esophagus is known as
    (a) Bronchography   (b) Myelography
    (c) **Esophagography**   (d) Dacrocystorhinography

33. Contrast radiography of renal architecture is known as
    (a) Bronchography   (b) Myelography
    (c) Esophagography   (d) **Intravenous pyelography**

34. Contrast radiography of architecture of spinal cord and its roots is known as
    (a) Bronchography   (b) **Myelography**
    (c) Esophagography   (d) Intravenous pyelography

35. Contrast agent skiodan and pantopaque are used for
    (a) Bronchography   (b) **Myelography**
    (c) Esophagography   (d) Intravenous pyelography

36. Contrast agent conray 420 is used for
    (a) Bronchography   (b) Myelography
    (c) Esophagography   (d) **Intravenous pyelography**

37. Contrast agent Dionosil is used for
    (a) **Bronchography**   (b) Myelography
    (c) Esophagography   (d) Intravenous pyelography

38. Contrast agent $BaSO_4$ is used for
    (a) Bronchography   (b) Myelography
    (c) **Esophagography**   (d) Intravenous pyelography

39. After $BaSO_4$ meal, radiograph is taken for oesophagus after
    (a) **Immediately**   (b) 30 minutes
    (c) 1 hours   (d) 3 hours

40. After BaSO$_4$ meal, radiograph is taken for reticulum after
   (a) Immediately
   (b) **30 minutes**
   (c) 1 hour
   (d) 3 hours

41. After barium meal, radiograph is taken for colon after
   (a) Immediately
   (b) 30 minutes
   (c) 1 hour
   (d) **3 hours**

42. After barium meal, radiograph is taken for rectum after
   (a) Immediately
   (b) 30 minutes
   (c) **5 hours**
   (d) 3 hours

43. Dose of BaSO$_4$ for large animals is
   (a) **½ kg to 1 kg**
   (b) 1 ounce / 5lb Body Weight
   (c) 0.3 ml / Kg Body Weight
   (d) 400 mg / lb Body Weight

44. Dose of BaSO$_4$ suspension for dog (canine) is
   (a) ½ kg to 1 kg
   (b) **1 ounce / 5lb Body Weight**
   (c) 0.3 ml / Kg Body Weight
   (d) 400 mg / lb Body Weight

45. For myelography dose of skiodan is
   (a) ½ kg to 1 kg
   (b) 1 ounce / 5lb Body Weight
   (c) **0.3 ml / kg Body Weight**
   (d) 400 mg / lb Body Weight

46. For urography dose of iodine in dog is
   (a) ½ kg to 1 kg
   (b) 1 ounce / 5lb Body Weight
   (c) 0.3 ml / Kg Body Weight
   (d) **400 mg / lb Body Weight**

47. Scatter radiation can be controlled by
   (a) Beam collimators
   (b) Grid
   (c) **Both**
   (d) None

48. Splash of fixer on the film before processing will produce
   (a) Black spot
   (b) **White spot**
   (c) Both
   (d) None

49. When the source of radiation is kept at a distance from the lesion, the radiation therapy is known as
   (a) Brachytherapy
   (b) Pliesotherapy
   (c) **Telethreapy**
   (d) Interstitial therapy

50. Difference in radiographic density between adjacent area is called as
   (a) Radiographic density
   (b) **Radiographic contrast**
   (c) Radiographic detail
   (d) None

51. Splash of developer on the film before processing will produce
   (a) **Black spot**
   (b) White spot
   (c) Both
   (d) None

52. When the source of radiation is kept within the interstitium, the radiation therapy is known as
    (a) **Brachytherapy**   (b) Pliesotherapy
    (c) Telethreapy   (d) Interstitial therapy

53. The maximum field of view which can be obtained with a specific radiographic system is generally limited by the
    (a) Anode angle   (b) Heel effect
    (c) Focal spot size   (d) **Both (a) and (b)**

54. The maximum mA which can be used for a single radiographic exposure is related to
    (a) Kv   (b) Exposure time
    (c) Focal spot size   (d) **All of the above**

55. Relatively low kV values are used in some X-ray procedures for the purpose of
    (a) Increasing penetration   (b) **Increasing contrast sensitivity**
    (c) Both (a) and (b)   (d) None

56. Changing the kV from 90 to 70 will generally
    (a) Require an increase in mAs by at least a factor of 2
    (b) Increase patient exposure
    (c) **Both (a) and (b)**
    (d) None

57 Changing from a 5:1 ratio to a 10:1 ratio grid will increase
    (a) Patient exposure   (b) Required kV or mAs
    (c) Image contrast   (d) **All of the above**

58. Changing from a 10:1 ratio to a 5:1 ratio grid
    (a) Patient exposure is decreased (b) Less grid cut off
    (c) **Both (a) and (b)**   (d) None

59. If grid is changed from low ratio to a high ratio you would expect
    (a) Patient exposure is decreased (b) **An increase in image contrast**
    (c) Both (a) and (b)   (d) None

58. The thickness of an intensifying screen has a significant effect on
    (a) Image blurring   (b) Patient exposure
    (c) **Both (a) and (b)**   (d) None

59. An underdeveloping of radiographic film can result in increased film
    (a) Contrast   (b) Fog
    (c) Density   (d) **None**

60. Condition which can reduce contrast in a general radiographic image include
    (a) Underexposure   (b) Overexposure
    (c) Overdevelopment   (d) **All of the above**

61. The amount of contrast on a radiograph can be affected by
   (a) Latitude of the film   (b) Processing condition
   (c) Amount of exposure   (d) **All of the above**

62. Potential advantage of using a higher kV (90 rather than 70) in radiography include
   (a) Reduced patient exposure   (b) Shorter exposure time
   (c) Reduced X-ray tube heating   (d) **All of the above**

63. When the smaller focal spot size of an X-ray tube is selected, you would expect:
   (a) Reduced scatter radiation   (b) **Improved detail**
   (c) Increased image noise   (d) All of the above

64. When using a magnification technique in radiography it is essential to have
   (a) Low kV   (b) Low mAs
   (c) A short exposure time   (d) **A small focal spot**

65. A small focal spot is used to
   (a) Reduce patient exposure   (b) Reduced image blurring
   (c) Increased visibility of detail (d) **Both (b) and (c)**

66. An air gap technique will generally improve image contrast because
   (a) Air absorbs scatter radiation
   (b) It is used with a small focal spot
   (c) **The scatter is more diverging than the primary beam**
   (d) None

67. The primary factor that limits the maximum mA that can be used during a radiographic exposure is
   (a) Anode angle   (b) **Focal spot size**
   (c) Cathode temperature   (d) Exposure time

68. If a radiographic procedure requires 20 mAs for a focal object distance (FOD) of 40 inch, a FOD of 80 inch would require
   (a) 5 mAs   (b) 10 mAs
   (c) 40 mAs   (d) **80 mAs**

69. Exposure time for radiography of avian and exotic animals should be
   (a) 1 second or less   (b) 1/20 second or less
   (c) 1/50 second or less   (d) **1/40 second or less**

70. Birds require smaller exposure factors than reptiles and mammals of same thickness because
   (a) Bird's cortices are thinner   (b) Bones have less calcium
   (c) **Both (a) and (b)**   (d) None of the above

71. How long should a bird be fasted before administration of contrast media for a gastrointestinal study

    (a) 12 hours   (b) 16 hours

    (c) **Not more than 4 hours**   (d) 8 hours

TRUE/ FALSE

(a) Radiation therapy is more effective in oxygenated cells. (**T** / F)

(b) For viewing the lateral view (L) of axial skeleton, cranial (rostral) aspect of the animal should be towards the radiologist's left. (**T** / F)

(c) In ultrasonography, highly reflective interfaces give hyperechoic image. (**T** / F)

(d) Pulse echo principal is used for ultrasound imaging. (**T** / F)

(e) Therapeutic ultrasound is indicated in synovitis and adhesions caused by trauma. (**T** / F)

(f) The emulsion layer of the X-ray film contains silver halide crystals. (**T** / F)

(g) Pregnant woman can be allowed to restrain the animal in the exposure room. (T / **F**)

(h) One radiographic view is sufficient for diagnosis of any deformity/lesion. (T / **F**)

(i) Intensifying screen increases the exposure factors requires to obtain a diagnostic radiograph. (T / **F**)

(j) Fluoroscopy permits clinical evaluation of the dynamics of the body. (**T** / F)

(k) Cholycystapaques are exclusively excreted through the biliary system. (**T**/ F)

(l) Use of diluted or exhausted fixer leads to yellow radiograph. (**T** / F)

(m) Nerve and muscles are relatively radioresistant. (**T** / F)

(n) Ultrasound therapy helps in resolution of chronic inflammatory processes. (**T** / F)

FILL IN THE BLANK WITH APPROPRIATE WORD(S)

- Contrast radiographic study of blood vessels is called as **Angiography**
- For routine radiography FFD is kept **90-100cm**.
- Degree of blackness on the processed film is known as **Radiographic density**
- For myelography, the contrast agent of choice is **Metrizamide**
- X-rays were discovered by **W.K. Roentgen,**.in the year **1895**
- Fixer contains the **Sodium thiosulfate**
- In B-mode of display, B stands for **Brightness**
- Father of veterinary radiology **R. Eberlein**
- Filament of X-ray tube is made up of **Tungsten**

- The radiation monitoring devices are **Film badge and Thermoluminescent dosimeter**
- Main scatter radiation control devices are **Collimator,** and **Grid**
- Contrast radiographic study of the nasolacrimal duct is known as **Dacrocystorhinography**
- Pyrimidines are more radiosensitive than **Purines**
- The average transit time of barium to travel from the gizzard to the cloaca in birds **½ to 4 hours.**

Absorbed dose: That amount of energy from ionizing radiations absorbed per unit mass of matter and expressed in rads.

Accelerators: Chemicals that increases the pH of the developer and subsequent increase the rate of developing.

Acidifiers: The agents that neutralize the alkaline developer and accelerate the fixing process.

Acoustic enhancement: Fluid gives an anechoic image as the sound beam passes uninterrupted. A normal bright area of increased sound intensity is produced beyond a fluid filled space. This phenomenon is called acoustic enhancement *e.g.* during normal scanning of urinary bladder.

Acoustic impedance: interference with the passage of sound waves by objects in the path of those waves. It equals the velocity of sound in a medium multiplied by the density of the medium. The acoustic impedance of bone may be nearly five times as great as that of blood.

Acoustic shadowing: The most common acoustic shadows of clinical significance are produced by cystic, renal and biliary calculi.

Actual focal spot: The actual area of the focal spot on the radiographic target as viewed at right angles to the plane of the target.

Afterglow: The tendency of a luminescent compound to continue to give off light after x-radiation has stopped.

Air gap technique: A technique to reduce scatter of radiation by increasing the distance between the patient and the surface of the cassette.

ALARA (As Low As Reasonably Achievable) principle: It is the official method of choice for limiting exposure to radiation.

ALARA MPD: The MPD for non-occupational person is as low as 0.1 rem/year or 1mSv/year.

Alloy: A mixture of metals. Anode of X-ray tube is made up of tungsten-rhenium alloy

A-mode: It represents the time required for the ultrasound beam to strike a tissue interface and return its signal to the transducer. The greater the reflection at the tissue interface, the larger the signal amplitude on the A-mode screen.

Anechoic: An absence of internal echoes. A cyst filled with clear fluid appears anechoic (black).

Angiocardiography: Sequential production of radiographs during the injection and circulation of the contrast agent through the heart and blood vessels.

Angiography: Angiography is the contrast radiographic study of the blood vessels. **Arteriography:** is a type of angiography that involves the study of the arteries.

Anode: the electrically positive terminal of a roentgen ray (radiographic) tube; a tungsten block embedded in a copper stem and set at an angle of 20° or 45° to the cathode. The anode emits roentgen rays (radiographs) from the point of impact of the electronic stream from the cathode.

Antegrade pyelography: In this technique the contrast medium is introduced into the renal pelvis by percutaneous needle puncture.

Arcing: The metal deposits on the inner wall of the envelope of X-ray tube act as a secondary anode. The electrons are attracted from the cathode towards the secondary anode instead of main anode. This phenomenon is called as arcing.

Arthrography: Contrast radiographic study of the articular surfaces and joint capsule after injecting contrast media (positive or negative) in to the joint space.

Artifacts: Anything that decreases the quality of the radiograph or ultrasound, CT, MRI images resulting in difficult evaluation and interpretation.

Attenuation: The energy loss in a beam of radiation or sound beam as it passes through matter. The intensity of the electromagnetic radiation decreases as its depth of penetration increases. A sound beam becomes weaker as it travels through tissue.

Automatic film processing: A fast and efficient method of processing in which the film is mechanically transferred from the developer to the fixer, is washed, and finally is dried.

Autoradiograph / Autoradiogram: An image recorded on a photographic film or plate produced by the radiation emitted from a specimen, such as a section of tissue, that has been treated or injected with a radioactively labeled isotope or that has absorbed or ingested such an isotope.

Autotransformer: It provides a variable yet predetermined voltage to the high voltage step up transformer.

Axial resolution: The ability of an ultrasound system to separate two objects lying along the axis of an ultrasound beam.

Back scatter: Backward diffusion of X-rays.

Barium Enema / lower GI (gastrointestinal) exam / lower bowel series: Contrast radiographic study of the colon / large intestine (Position and contour). A dilute (5 to 20%) suspension of barium is introduced into emptied colon.

Barium meal: A strong (usually 100%) suspension of barium sulfate is administered to an animal which has been starved for at least 12 hours.

Barium study: contrast radiographic examination using a barium mixture to locate disorders in the esophagus, stomach, duodenum, and the small and large intestines.

Barium sulfate: A water-insoluble salt used as an opaque contrast medium for radiographic examination of the digestive tract.

Barium swallow: Contrast radiographic or fluoroscopic examination of swallowing and esophageal function after oral administration of barium paste or liquid.

Barium-impregnated polyethylene spheres (BIPS): Radio-opaque markers used to demonstrate intestinal obstruction and motility disorders; the spheres are given orally and their movement can be tracked radiographically.

Black spots: The spots caused by dust particles or developer on the films before development; also caused by outdated (expired) film.

Blurred image: Caused by film movement during exposure, bent film during exposure, double exposures, or flowing of emulsion during processing in excessively warm solution.

B-mode: Brightness modulation in diagnostic ultrasonography. Bright dots on an oscilloscope screen represent echoes, and the intensity of the brightness indicates the strength of the echo.

Brachytherapy: Radiotherapy in which the radiation source (iridium-192, radium-226) is applied to the surface of the body or within the body a short distance from the area being treated.

Brain angiography: Radiography of the cranium after the intravenous injection of a radiopaque substance. An area of poor vascularity indicates the presence of a space-occupying lesion in the brain.

Bremsstrahlung or braking radiation: The high speed electron passes close to the nucleus of the target atom (attracted by the positive charge) and slows as it binds towards the nucleus and releases energy in the form of electromagnetic radiation called Bremsstrahlung or braking radiation.

Bucky diaphragm /Bucky grid: A moving grid that limits the amount of scattered radiation reaching a radiographic film, thereby increasing the film contrast.

Caliper: It is used to measure the thickness of anatomic parts.

Cassette grid: Composed of alternating strips of lead and radiolucent material such as aluminum. Placed on top of the cassette it permits the passage only of the X-rays that are passing directly to the film.

Cassette holder: A holder having a long handle, into which the cassette fits. It protects the person holding the cassette from the primary X-ray beam.

Cassette, cardboard or non-screen type: A cardboard envelope of simple construction suitable for use in making radiographs.

Cassette, screen-type: A cassette usually made of metal, with the exposure side of low-atomic-number material, such as Bakelite, aluminum, or magnesium, and containing intensifying screens between which a "screen type" of film or films may be placed for exposure to radiographs.

Cassette: A light-proof housing for X-ray film, containing front and back intensifying screens, between which the film is placed.

Cathode ray tube (CRT): A vacuum tube in which a beam of electrons is focused to a small point on a luminescent screen.

Cathode: The negative side of the X-ray tube, which consists of the molybdenum focusing cup and a helical tungsten filament. It provides a source of electron.

Caudal: The part of the head, neck and trunk positioned towards the tail from any given point. It also describes aspect of the limb facing tail but proximal to the carpal and tarsal joints.

Celiography: Contrast radiographic study for evaluating the abdominal cavity and diaphragm.

Cerebral angiography: Radiographic procedure used to visualize the vascular system of the brain after injection of a radiopaque contrast medium.

Cerebral angiography: Used to detect vascular irregularities in the brain.

Cholecystography: Contrast radiographic study of gall bladder and bile ducts.

Clear radiograph: The result of treating the film with fixer before developing or by excessive washing. The problem can be prevented by following appropriate procedures.

Clearing agents: See fixing agents

Collimator / Diaphragm: Control the direction and the dimensions or limits the size and shape of the primary X-ray beam. It is used to reduce scatter radiation, thereby decreasing the patient dose needed and increasing radiographic quality.

Comet tail: Caused by highly reflective interfaces specially the air fluid interface *e.g.* In partially consolidated lung at the Interface between diaphragm and lung. Interface between the bowel wall and bowel gas.

Compton effect: Collision between a photon and a particle results in an increase in the kinetic energy of the particle and a corresponding increase in the wavelength of the photon.

Console: The control panel of the X-ray machine.

Contrast media: A substance that is either radiolucent or radiopaque and can be administered to increase radiographic contrast within an organ or system.

Coupling gel: A water soluble gel which is applied to the skin to eliminate an air interface and permit transmission of the ultrasound beam from the transducer into the body.

Cranial sinus venography: Radiographic visualization of structures of the cranial vault. It can be divided into dorsal sagittal sinus venography and cavernous sinus venography.

Cranial: The part of the neck, trunk and tail positioned towards the head from any given point. It also describes aspect of the limb facing the head above carpal and tarsal joints.

Crisscross grid: See Crossed grid.

Cross-hatched grid: Two linear grids that are superimposed at right angles to each other and used for maximal removal of scatter radiation.

CT number: The density of each voxel is compared with density of water and then assigned to a gray scale shed. It represents the attenuation of the X-ray beam in tissue within a voxel. CT number for Metal (+3000), Bone (+1000), Air (-1000) and Water: (0).

Cystography: The radiographic examination of the urinary bladder after introduction of a contrast medium.

Dacrocystorhinography: Contrast radiographic examination of nasolacrimal duct.

Dark radiograph: Caused by overexposure of the film to radiation, film fog from extended development, accidental exposure to light (light leaks in film packet or dark room), or an unsafe darkroom light.

Dental film: Non-screen film used in dental radiography.

Developer: A chemical solution that converts the invisible (latent) image on a film into a visible one by reducing the light-activated silver halide molecules to metallic silver. A standard developer is a mixture of Metol and Hydroquinone.

Developing agents: The agents used to convert a latent image on exposed X-ray film to a visible image.

Diffusion MRI: It measures the diffusion of water molecules in biological tissues. In **diffusion tensor imaging (DTI)** diffusion is measured in multiple directions.

Discography: Examination of the intervertebral disk space using X-rays after injection of contrast media into the disk.

Distal (Di): Situated away from the point of attachment or origin.

Distant enhancement: See Acoustic enhancement.

Doppler echocardiography: In doppler echocardiography, ultrasound is directed towards the heart and is reflected by R.B.Cs. The frequency of reflected signals is altered in relation to the velocity of RBCs. Direction and speed of the moving target can be calculated by the change in frequency of the Doppler signal.

Doppler shift: Difference between transmitted and received sound frequencies is known as Doppler shift.

Dorsal: Towards the vertebrae. Upper aspect of the head, neck, trunk and tail and dorsal aspect of the limb from the carpal or tarsal joint distally.

Dose equivalent (DE): Dose equivalent expresses the amount of radiation dose and the physical damage that it may produce. It is the product of the dose (in rad or gray) and a quality factor specific to the type and energy of the radiation delivering that dose. The unit of dose equivalent is the sievert (Sv) -SI unit or the rem.

Dosimeter: A device used to measure radiation exposure to personnel.

Dosimetry: Determination of the amount, rate and distribution of radiation, especially X-rays or gamma rays to which an animal or person has been exposed during a given period.

Double contrast gastrography: Radiography of the stomach taken after administration of barium and then air. This technique is useful in evaluating details of the gastric mucosa.

Double contrast: A radiographic contrast technique that uses a combination of positive and negative contrast media simultaneously.

Dyschroic fog: Fogging of the radiograph, characterized by the appearance of a pink surface when the film is viewed by transmitted light and a green surface when the film is seen by reflected light. It usually is caused by an exhaustion of the acid content of the fixing solution (incomplete fixation).

Echocardiography: An echocardiogram is the sonogram of the heart.

Echogenic: Structures that reflect high-frequency sound waves and thus can be imaged by ultrasound techniques.

Edge shadow / refraction: Artifact occurs when the incident beam interacts with a curved surface *e.g.* gall bladder, kidney, cysts, fetal skull etc. The sound beam is bent and diverges resulting in a narrow or broad shadow deep to the curved surface.

Effective focal spot: The area of the focal spot that is visible through the X-ray tube window and directed towards the X-ray film.

Electroagnetic radiation: A method of transporting energy through space distinguished by wavelength, frequency and energy.

Electron volt (eV) and Kilo electron volt (keV): Unit for measurement of energy of an X-rays.

Electron: It is a negatively charged particle that travels around the nucleus.

Elongation: The image appears longer than actual size because X-ray beam is not directed perpendicular to the film surface.

Epiduralograms: Contrast radiographic study of epidural space after injecting air or positive contrast agent.

Esophagography: Radiographic visualization of the esophagus using a swallowed radiopaque contrast medium.

Eulsion: A layer of radiographic film which contains finely precipitated silver halide crystals in a gelatin base.

Excretory urography: See Intravenous pyelography.

Exposure time: The period of time during which X-rays are permitted to leave the X-ray tube.

External beam radiotherapy: see Teletherapy.

Fasciagraphy: Contrast radiographic study of tendons and associated structures.

Filament: Consists of the tungsten and emits the electrons. It is a part of low energy circuit in the cathode.

Film badge: A lightproof pack of radiographic film, usually worn on the body during exposure to radiation in order to detect and quantitate the dosage of exposure.

Film base: A transparent flexible polyester or cellulose acetate support layer of the radiographic film ad gives blue tint to the film.

Film hanger: An instrument or device for holding radiographic film during processing procedures.

Film latitude: The exposure range of a film that will produce acceptable densities.

Film marker: Lead letters are placed on the film loaded cassette to indicate which part of the animal was examined and the projection used.

Film processing: A chemical transformation of the latent image, produced in a film emulsion by exposure to radiation, into a stable image visible by transmitted light. The usual procedure is basically a selective reduction of affected silver halide salts to metallic silver grains (development), followed by the selective removal of unaffected silver halide (fixation), washing to remove the processing chemicals, and drying.

Film speed (film sensitivity): The amount of exposure to light or roentgen rays required to produce a given image density. It is expressed as the reciprocal of the exposure in roentgens necessary to produce a density of 1 above base and fog; films are classified on this basis in six speed groups, between each of which is a twofold increase in film speed.

Fistulography: Contrast radiographic study of draining tracts like sinus and fistula to determine the depth and origin of tract.

Fixation: It is the process of removal of unexposed silver halide crystals from the film and the gelatin is hardened.

Fixer: The agents used for fixation.

Fixing agents: These agents dissolve and remove the unexposed silver halide crystals to black metallic silver.

Flat film: A film lacking in radiographic contrast.

Fluorescein angiography: Fluorescein angiography is used to diagnose retinal problems and circulatory disorders.

Fluoroscence: The ability of a substance to emit visible light.

Fluoroscopy or radioscopy: Fluoroscopy is the dynamic radiological study of the body parts. The technique offers continuous imaging of the motion of internal structures and immediate serial images.

Focal spot: The focal spot is the area of anode (target) which is bombarded by electrons from the cathode. The focal spot is oriented at 11^0 to 20^0 angles.

Focal-film distance: the distance between the anode of the X-ray tube and the film; an important exposure value.

Focused grid: A linear grid in which the lead strips are aligned in a progressively tilted fashion toward a centering point.

Focusing cup: Consists of the molybdenum and focus the electron emission toward the target of the anode.

Fog: Caused by stray radiation, use of expired film, or an unsafe darkroom light.

Foreshortening: Distortion of the anatomic structures when the image appears shorter than the actual object. It is caused by excessive vertical angulation.

Full wave rectification: Creates an almost constant electrical potential across the X-ray tube, converting the positive electrical current pulses to 120 times / second.

Functional MRI (fMRI): It measures the hemodynamic response related to neural activity in the brain and spinal cord. It is best for neuro-imaging.

Gamma camera: It consists of a rare earth activated sodium iodide crystal that is tightly sealed to an array of many closely aligned "light pipes". Gamma camera is the most common radiation detector in veterinary use.

Gamma rays: The electromagnetic radiation emitted from the nucleus of radioactive substances.

Gantry: It is a ring that contains X-ray tube that is positioned on the opposite side of the detectors. It can be moved 360^0 around the patient.

Gastrography: contrast radiography of the stomach.

Genetic damage: Effects of radiation that occur to the genes of reproductive cells.

Geometric unsharpness: Loss of radiographic detail due to geometric distortion.

Glass envelope: A glass vacuum tube that contains the cathode and anode of the X-ray tube.

Gray (Gy): The amount of radiation such that the absorbed energy is 1 joule/kg of tissue.

Grid cutoff: Differences in radiographic intensity that are caused by improper focusing of the lead lines of a grid and subsequent absorption of primary X-rays by the grid lines.

Grid efficiency: The ability of a grid to absorb scatter radiation in the production of quality radiograph.

Grid factor: The amount of exposure needs to be increased to compensate for the grid's absorption of a portion of the primary beam.

Grid focus: The distance between the focal spot and grid in which the grid is effective without grid cutoff.

Grid ratio: Ratio of the height of the lead strips to the width between two lead strips.

Grids: A grid is a series of thin, linear strips of alternating radio-dense material (lead) and radiolucent interspaces (plastic, aluminum or fiber). The grids are used to reduce the amount of scattered radiation reaching the X-ray film.

Half wave rectification: A method of converting AC to DC in which half of the current is lost.

Hardeners: The agents added to the fixing solution to prevent excessive emulsion swelling.

Heel effect: The X-ray intensity greater at the cathode end of the X-ray field and lower at the anode end because of absorption in the target material.

Hounsfield number: See CT number.

Hyperechoic: Echoes produced are brighter than in surrounding tissue and characteristic of bone and dense tumor tissue.

Hypoechoic: Tissues or structures that reflect relatively few of the ultrasound waves directed at them.

Image intensifier: X-ray image intensifier converts low intensity X-rays in to visible image.

Intensifying screens: These screens contain luminescent phosphor crystals. The crystals emit foci of light when exposed to X-rays.

Internal beam radiation therapy: See Brachytherapy

Interstitial brachytherapy (IB) or Curie therapy: Radiation source is implanted directly into the tumor in the form of small pellets, grains, seeds, wires, tubes.

Interventional MRI: Images produced by MRI are used to guide minimally invasive procedures.

Intravenous phlebography: See Osteomedullography

Intravenous pyelography (IVP): Contrast radiographic technique for examining the structure and function of the kidneys and ureters. A contrast medium is injected intravenously, and serial X-ray films are taken as the medium is cleared from the blood by the kidneys. The renal calyces, renal pelvis, ureters, and urinary bladder are all visible on the radiographs.

Intravenous urography: See Intravenous pyelography.

Inverse square law: The intensity of the radiation varies inversely as the square of the distance from the source.

Ion chambers: These devices are charged before being worn. After each exposure, the ion chamber discharges. The amount of discharge of the ions is proportional to the amount of radiation received.

Isoechoic: Returning waves of normal amplitude in ultrasonography.

Kilovoltage peak (kVp): The peak energy of the X-rays which determines the quality (penetrating power) of the X-ray beam.

Kilovoltage: The amount of electrical energy being applied to the anode and cathode to accelerate the electrons from the cathode to the anode.

Labelled compound: The compound tagged with radionuclide.

Larmor frequency: The frequency of the precession of a charged particle when its motion comes under the influence of an applied magnetic field and a central force.

Latent image: An invisible image on the X-ray film after it is exposed to ionizing radiation or light before processing.

Light radiograph: Caused by under exposure, underd evelopment (expired or diluted developing solution), and development in temperatures that are too cold or accidental use of a wrong film speed.

Line focus principle: The effect of making the actual focal spot size appear smaller when viewed from the position of the film because of the angle of the target to the electron stream.

Linear array probe: Ultrasound probe which contains multiple in-line transducers that crearte a rectangular shaped image.

Linear energy transfer (LET): The amount of energy deposited as the particles traverse a section of tissue is referred as LET.

Linear grid: A grid designed to permit the passage of the primary beam by having lead lines aligned in the same direction separated by radiolucent interspacing material. There are two types, parallel and focused.

Lines per centimeter: The number of lead stripes per centimeter area of a grid.

Lower gastrointestinal (LGI) study: A radiographic contrast study evaluating the caecum colon and rectum.

Lymphography: Contrast radiographic study of lymph nodes and lymphatic vessels.

Magnetic resonance angiography: It is used to generate pictures of the arteries to evaluate stenosis or aneurysms. Magnetic resonance venography (MRV) is used for veins.

Magnetic resonance spectroscopy (MRS): It provides the chemical information of that region. MRS detects the intracellular relationship of brain metabolites.

Magnification: Distortion of anatomic structures when the image appears larger than actual size.

Main bang: Near field artifact created by high level echoes at the skin's surface.

Maximum permissible dose (MPD): The maximum dose of radiation a person may receive in a given time period. MPD (annual)= 5 rems (N-18) where N is the age in years.

Milliamperage (mA): The amount of electrical energy being applied to the filament. Milliamperage describes the number of X-ray produced during the exposure.

Milliamperage-seconds (mAs): The number of X-rays produced over a given period.

Mirror image: Highly reflecting interfaces act like a mirror and shows mirror image. This defect is common during scanning of tendons.

M-mode / TM-mode: A diagnostic ultrasound presentation of the temporal changes in echoes in which information is displayed as depth versus time on a graph. Used for echocardiography.

Myelography: Contrast radiographic evaluation of the spinal cord after injecting positive contrast agent into the subarachnoid space.

Negative contrast agents: These agents are more radiolucent to X-rays and have a black appearance on a finished radiograph.

Non screen films: The film that is more sensitive to ionizing radiation than to fluorescent light.

Nonselective angiography: Injection of contrast material into a regional vessel or the general circulation.

Object-film distance: The distance, usually expressed in centimeters or inches, between the object being radiographed and the cassette or film.

Optically stimulated luminescence (OSL) badge: A thin layer of aluminum oxide is stimulated with different frequencies of Laser light. Illumination of aluminum oxide is proportional to the amount of radiation received. OSL badge are extremely sensitive to the radiation.

Orbital angiography: Contrast radiographic study of the arteries of the orbit.

Osteomedullography: Contrast radiographic study of the intraosseous and extraosseous venous channels of the long bones.

Palmer (Pa): Situated on the caudal aspect of the front limb, distal to the carpal joint.

Parallel grid: A grid with lead stripes that are parallel and at right angles to the film.

Penumbra: The region of blurred edge to an image, a halo effect, in an X-ray film caused usually by an overlarge focal spot exacerbated by a long object-to-film distance. It is due to geometric unsharpness.

Photoelectric effect: The ejection of electrons from a solid by an incident beam of sufficiently energetic electromagnetic radiation.

Photon: A bundle of radiant energy.

Piezoelectric crystal: The crystals present in transducer vibrate and produces sound wave. It is made up of Lead ZirconateTitanate (PZT).

Pixel: A picture element. Tiny squares make up the image matrix. The two dimensional image that is produced is composed of many squares called pixels. It is the smallest image-forming unit of a video display. A CT or PET scan is composed of an array of squares (pixels), each of which is coloured a uniform shade of gray or another colour. The greater the concentration of pixels the clearer the image. The CT or MR image, usually have 512 × 512 or 256 × 256 pixels, respectively.

Planter (Pl): Situated on the caudal aspect of the rear limb, distal to the tarsal joint.

Plesiotherapy (Surface therapy): Radiation source is applied on to the tumor surface.

Pneumocystogram: Negative contrast radiographic technique for evaluating the urinary bladder.

Pneumoencephalography: Radiographic visualization of the ventricular space, basal cisterns, and subarachnoid space overlying the cerebral hemispheres of the brain. Air, helium, or oxygen is injected into the lumbar subarachnoid space after the intermittent removal of the cerebrospinal fluid by lumbar puncture.

Pneumopericardiography: It is the negative contrast radiographic examination of pericardial sac.

Pneumoperitoneography: It is the negative contrast study of the abdominal cavity after injection of air/gas into the peritoneal space for increasing the subject contrast.

Pocket ionization chamber: See ionization chamber.

Positive contrast agents: These agents are more radio-opaque to X-rays and have a white appearance on a finished radiograph.

Positron Emission Tomography (PET) PET is a nuclear medicine imaging technique. It produces a 3 D image of functional processes in the body.

Potter-Bucky grid: A mechanical device that consists of a focused grid within a diaphragm, which moves the grid across the X-ray beam during exposure so that no lines appear in the radiograph. It prevents scattered radiation from reaching the film, thereby securing better contrast and definition.

Preservatives: Agents that prevent rapid decomposition of the developer or fixer.

Primary beam: The path of X-rays follows as they leave the X-ray tube.

Proximal (Pr): Situated closer to the point of attachment or origin.

Pseudofocused grid: A grid with parallel lead stripes that are progressively reduced in height towards the edges of the grid.

Pulmonary angiography: Contrast radiographic study of pulmonary veins and arteries by introduction of contrast material into the jugular or cephalic vein or via a catheter positioned in the pulmonary artery.

Pulse echo principal: The short pulses of ultrasound are emitted and spaced far enough apart in time to give distant echoes enough time to return to the transducer before the next pulse.

Pyelography / Ureteropyelography / Pyeloureterography: Contrast radiographic of the renal pelvis and ureter (kidney and associated structures).

Quanta: See photon.

Quantum mottle: Quantum mottle is a density variation on finished radiograph due to random spatial distribution of phosphor crystals within the intensifying screen.

Radiation absorbed dose (Rad): It is the unit of absorbed dose following exposure to any ionizing radiation.

Radiation monitoring devices: Radiation monitoring devices are used to record the amount of radiation received by the person during radiographic procedures.

Radiation therapy: Treatment of diseases or solid tumors and occasionally that of benign conditions with the use of ionizing radiation.

Radiation tolerance: There is a radiation dose limit at which normal tissues are irreparably damaged.

Radiocurability: A radiocurable tumor is one that can be destroyed by a dose of radiation that is well tolerated by surrounding normal tissues.

Radiograph: A visible photographic record on film produced by X-rays passing through an object.

Radiographer: A technically trained person who can obtain quality radiograph for use by a radiologist.

Radiographic contrast: Difference in radiographic density between adjacent areas on a radiograph is known as radiographic contrast.

Radiographic density: Degree of blackness on the radiograph produced by the interaction of silver halide crystals with developing agents.

Radiographic detail: Radiographic detail is characterized by sharp tissue and organ interfaces.

Radiographic mottle: Density variation in a radiograph made with intensifying, given a uniform exposure. It is the result of quantum mottle, structure mottle and film graininess.

Radiographic or film fault: A defective result in a radiograph; usually caused by a chemical, physical, or electrical error in its production.

Radiography: The production of photographic images on film using radiation from other radioactive substances instead of light. Also called **X-ray scotography, shadowgraphy**

Radioisotopes: A radioactive isotope; one having an unstable nucleus and emitting characteristic radiation during its decay to a stable form. These radioisotopes are used to detect (diagnosis) and to treat the abnormality (therapy).

Radiologist: The person who is specialist in interpretation of the recorded radiograph.

Radiolucent: Permitting the passage of radiant energy, such as X-rays, with little attenuation. The representative areas appeared dark on the exposed film. A negative contrast agent produces radiolucent area on the finished radiograph.

Radionuclide: A nuclide that exhibits radioactivity. Radionuclides are combined with other chemical compounds or pharmaceuticals to form radio-pharmaceuticals.

Radiopaque: Not allowing the passage of X-rays or other radiation. The representative areas appeared white on the exposed film when a positive contrast agent is used.

Radiopharmaceutical: These are the drugs containing radioactive substances and used for diagnostic and therapeutic purposes.

Radioprotectors: Substances that protects normal cells from radiation *e.g. Amifostine.*

Radiosensitibity: Risk of damage of neoplastic cell to absorbed radiation.

Radiosensitizers: Drugs that make cancer cells more sensitive to radiation are called radiosensitizers *e.g.* 5-fluorouracil (5-FU), Misonidazole, Nimorazole, Cisplatin and Cetuximab.

Real time MRI: It refers to continuous monitoring (filming) of moving objects in real time.

Rectification: Process of changing alternating current (AC) to direct current (DC).

Reflective layer: A layer on an intensifying screen that reflects from the phosphor layer towards the film.

Renal angiography: Contrast radiographic study of renal blood flow.

Replenisher: Strong solution of original developer to replace lost volume during developing process.

Restrainers: These agents limit the action of the developing solution to the exposed silver bromide crystals in the film *e.g.* KCl and KBr. These agents are also used as antifoggant.

Reticulation: A network of corrugations produced because of an excessive difference in temperature between darkroom solutions. This artifact is due to irregular expansion and contraction of the film emulsion resulting in a mottled density appearance.

Reticulography: Contrast radiographic study of reticulum after administration of positive contrast agents.

Retrograde pyelography: Pyelography after introduction of contrast material through the ureter.

Reverberation: Process of echo bouncing back and forth between the two interfaces is known as reverberation. The presence of air between the probe and the patient leads to reverberation artifact.

Ring down: It is a reverberation artifact and appears as a long series of closely spaced echoes and is caused by a very strong interface such as gas bubbles or metal.

Rinsing: A process of removal of excess developer solution before the film is placed in the fixer solution. The rinsing solution is usually water.

Roentgen (R): It is the quantity of an X-rays or gamma radiation which produces one electrostatic unit (2.08×10^9 ion pairs/cm^3) in 1 cc of dry air after its ionization at 0^0C and 760 mm Hg.

Roentgen equivalent man (rem): It is the amount used to express the dose equivalent that results from exposure to ionizing radiation. (1 m rem = 0.001 rem).

Roentgenogram: See Radiograph.

Roller marks: The dark lines on films caused by contaminated chemicals in automatic film processing units. This artifact can be prevented by cleaning and replenishing the developer and fixer solutions regularly.

Rostral (R): Area on the head situated towards the nose.

Rotating anode: An X-ray tube in which the anode rotates on an axis to increase X-ray production while dissipating heat.

Santes' rule: A method of estimating kilovoltage in relation to area thickness: (2 x thickness in cm) + 40 = kVp.

Scatter radiation: It is caused by interaction of the primary beam with objects in its path.

Scitigraphy: Formation of a two dimensional image from light flashes as a result of the interaction of energy with an absorbing material (scintillation). In scintigraphy, radio-isotopes are taken internally and the emitted radiation is captured by gamma camera to form a two dimensional image.

Screen film: Film with silver crystals is more sensitive to fluorescent light emitted from intensifying screens than to ionizing radiation.

Sealed source radiotherapy: See Brachytherapy.

Secondary radiation: See Scatter radiation.

Sector probe: Ultrasound probe having multiple rotating or oscillating transducers that produce a wedge shaped image.

Selective angiography: Placement of the catheter in the vessel or heart chamber being studied in order to provide the best possible contrast study of the suspected lesion.

Short axis view: Echocardiographic image showing the heart in transverse plane.

Short distance therapy: See Brachytherapy.

Sialography: Contrast radiographic examination of salivary glands and its ducts.

Side lobe artifact: It refers to lateral displacement of structures produced by minor beams of sounds travelling out in directions different from the primary ultrasound beam. Curved surfaces like diaphragm, gall bladder, highly reflective surface etc. are common example of this artifact.

Sievert (Sv): In SI system the unit of dose equivalency is sievert.

Silhouette effect/ border effacement: When two structures of the same radio-opacity are in contact with each other, their individual margins at the point of contact cannot be distinguished (*e.g.* Liver and stomach are in close contact) and composite shadow is formed on a radiograph.

Silver halide: A compound of silver bromide (90-99%) and silver iodide crystals (1-10%).

Skiagram: See Radiograph

Slice thickness artifact: This artifact mimics the presence of sediment in the gall bladder or urinary bladder (pseudosludge- has a curved interface). True sediment usually has a flat interface and settles at the dependent portion of the organ while changing the position of the animal.

Somatic damage: Damage to the body induced by radiation that is manifested in the lifetime of the recipient.

Sonographer: The skilled person for performing ultrasound examinations.

Sonologist: The person who is specialist in interpretation of the recorded ultrasound image.

Source image distance: See focal film distance.

Static electricity: An image in the emulsion that has the appearance of lightning and is caused by rapid opening of the film pocket or transfer of static electricity from the technician to the film.

Stationary anode: A non-rotating anode in an X-ray tube so that the target surface is comparatively small. This type of anode is usually fitted in dental and small portable radiography units.

Step down transformer: It reduces the X-ray machine input voltage from 110 or 220V to 10V to prevent burnout of the cathode filament.

Step up transformer: It increases the incoming voltage of 110 or 220V to thousands of volts.

Stop bath: A solution of acetic acid and water used to stop the development of the X-ray film by rapidly neutralizing the alkaline developer solution.

Structure mottle: Fluctuation in the density due to non-uniform structure of intensifying screen.

Subject contrast: The difference in density and mass of two adjacent anatomic structures.

Substraction technique: This is a photographic method for eliminating certain unwanted shadows from a radiographic film.

Summation shadows: It results when parts of a patient or object in different planes (*i.e.* not in contact with each other) are superimposed.

Supercoat: A clear protective layer on radiographic film.

Target: see the anode.

Teletherapy (External beam radiation therapy or EBRT or XRT): The source of radiation is outside the body at some distance (80-100 cm) from the target tissue.

Thermionic emission: The process of releasing electrons from their atomic orbits by heat.

Thermo luminescent dosimeters (TLDs): A method of dosimetry consisting of a chamber containing lithium fluoride or calcium fluoride crystals. These crystals store the energy produced by radiation exposure.

Timer switch: It controls the length of exposure.

Tomogram: Image produced by tomography

Tomography: Imaging by section.

Transrectal probe: The transducer is introduced in to the rectum for examination of urinary bladder, prostate gland and rectum.

Unfocused grid: See Parallel grid.

Upper GI series: Contrast radiographic study of the upper gastrointestinal tract including the esophagus, stomach, and duodenum after oral administration of barium paste or liquid.

Urethrography: Contrast radiographic study of urethra.

Urography: Radiography of the urinary tract.

Vaginography: Contrast radiographic study of female reproductive tract after administration of contrast agent into the vagina.

Valve tubes: Allows the flow of electrons in one direction only and called as self-rectifiers.

Ventral: Lower aspect of the head, neck, trunk and tail.

Ventriculography: Contrast radiographic study of ventricles of brain after the ventricular fluid is replaced by air or by an opaque medium.

Vertebral venography: Injection of contrast medium into the saphenous vein while the caudal vena cava is compressed causes the vertebral veins to be outlined; used to demonstrate cord compression.

Voxels/volume elements: Three dimensional box represented on an image matrix by the two dimensional pixels.

White spots: A fault caused by air bubbles clinging to the emulsion during development or by fixing solution spotted on the emulsion before development.

Windowing: CT produces a volume of data which can be manipulated through windowing in order to demonstrate various structures based on their ability to block the X-ray beam.

Xeroradiography: Picture of the body parts is recorded on paper rather than on film. A latent electrostatic pattern is produced on the surface of the photoconductor (amorphous selenium).

X-ray beam: A number of X-rays travelling together through space at a rapid speed.

X-ray tube: A vacuum tube containing cathode and anode that produces a controlled X-ray beam.

X-rays: A form of electromagnetic radiation similar to visible light but of a shorter wavelength.